Practical Biochemistry
An Introductory Course

Fiona Frais

BUTTERWORTHS LONDON
UNIVERSITY PARK PRESS BALTIMORE

THE BUTTERWORTH GROUP

ENGLAND

Butterworth & Co (Publishers) Ltd
London: 88 Kingsway, WC2B 6AB

AUSTRALIA

Butterworth & Co (Australia) Ltd
Sydney: 586 Pacific Highway, Chatswood, NSW 2067
Melbourne: 343 Little Collins Street, 3000
Brisbane: 240 Queen Street, 4000

CANADA

Butterworth & Co (Canada) Ltd
Toronto: 14 Curity Avenue, 374

NEW ZEALAND

Butterworth & Co (New Zealand) Ltd
Wellington: 26-28 Waring Taylor Street, 1

SOUTH AFRICA

Butterworth & Co (South Africa) (Pty) Ltd
Durban: 152-154 Gale Street

© Butterworth & Co (Publishers) Ltd, 1972

ISBN 0 408 70244 3 Standard
     0 408 70245 1 Limp

Published in 1972 jointly by
BUTTERWORTH & CO (PUBLISHERS) LTD, LONDON
and
UNIVERSITY PARK PRESS, BALTIMORE

Library of Congress Cataloging in Publication Data
Frais, Fiona.
  Practical biochemistry.
  Bibliography: p.
  1. Biological chemistry—Laboratory manuals.
I. Title.
QD415.5.F7   547'.7'028   72-471
ISBN 0-8391-0635-1

Printed in Hungary

Preface

I have intended that this book be read as a first-year practical course in biochemistry or applied biology. The book assumes that the student has a good knowledge of chemistry. It will be suitable for biology and applied biology courses at a variety of levels up to and including first-year undergraduate courses in biological subjects.

I have included a theoretical summary on the structure and reactions of each group of biological compounds at the beginning of each chapter and preceding each experiment. It is hoped that this will enable the student to understand fully all the experi ments.

The majority of the experiments can be completed in a three-hour practical session. In addition, most of the apparatus is neither highly specialised nor expensive. Experiments such as separation by gas chromatography, which require specialised apparatus, have been omitted.

The experiments included in this book are intended to demonstrate the properties of the main groups of biological compounds and to demonstrate many of the techniques used in their purification and separation. The properties of certain groups of compounds, e.g. nucleic acids, vitamins and hormones, have been omitted. There are several reasons for these omissions: in the case of vitamins and hormones, they act as a group physiologically but chemically they are very diverse and therefore one experiment is only characteristic of one compound. Moreover, the background chemistry of these compounds is considered to be outside the scope of this book. The chemistry of nucleic acids is limited and therefore experiments in this group involve extraction and separation; similar techniques have been explained in other sections of the text. Also, I consider that the methods required for the extraction of these compounds are too protracted to be included amongst the routine experiments described here.

I have included a section on statistics and graphs in the introductory chapter, since I believe that an understanding of these two

PREFACE

subjects forms an increasingly important part of a student's ability to present results in the most meaningful and effective way.

The chemicals required for each practical exercise are listed in each case, and the recipes for making up specific reagents—marked with an asterisk in the text—are given in the appendix.

I would add that I have taken great care to arrange the treatment of experiments in such a way that the content can be easily understood and absorbed by all students of biochemistry and biology, provided, as I have already said, that they have reached a reasonable level in the study of chemistry.

It is impossible to acknowledge the sources of all the material included and to thank all the people who have assisted me and so made this book possible. However, I should like to thank the following, who have given me particular assistance:

The Biochemistry Department of Cambridge University, Paddington Technical College and the Department of Biochemistry at the National Hospital, Queen Square, who afforded me time and facilities to carry out some of the experiments tested in this book. I should also like to thank J. Kinderlerer, who has helped me considerably with scientific content and advice on particular experiments, and K. N. Cross, Dr. Ann Williamson and Miss J. Lord, who have also given me valuable assistance. Finally I should like to thank Helen Robertson for typing the manuscript and my husband for spending long hours proof-reading the book.

As well as those defined in context, the following abbreviations are used:

ADP	Adenosine diphosphate
ATP	Adenosine triphosphate
BuOH	n-butanol
FAD	Flavin adenine dinucleotide
FMN	Flavin mononucleotide
HAc	Acetic acid
NAD	Nicotinamide adenine dinucleotide
NADP	Nicotinamide adenine dinucleotide phosphate
PhOH	Phenol

F. F.

Contents

Chapter 1 Introduction   1

Errors   1
Measurement of biological data   4
Graphs   6

Chapter 2 Carbohydrates   13

Structure of carbohydrates   13
Reactions of sugars   20

Experiment 1	Qualitative tests for carbohydrates	23
Experiment 2	Estimation of glucose by iodine oxidation in alkaline conditions	27
Experiment 3	Sugar content of fruit	29
Experiment 4	Estimation of sugars by Benedict's method	30
Experiment 5	Photometry: Colorimetric estimation of sugars	32
Experiment 6	Estimation of lactose in milk	37
Experiment 7	Polarimetry	39
Experiment 8	Periodate oxidation	42
Experiment 9	Glycogen levels in rat livers	46
Experiment 10	Separation of sugars by paper chromatography	49

Chapter 3 Amino acids and proteins   53

Structure of amino acids   53
Properties and reactions of amino acids   56
Protein structure   58
Denaturation   63
Precipitation of proteins   64
Separation and purification of proteins   65

CONTENTS

Experiment 11	Properties of amino acids	70
Experiment 12	Properties of proteins (serum proteins)	73
Experiment 13	Separation of amino acids by ion exchange chromatography	76
Experiment 14	Separation of amino acids by electrophoresis	79
Experiment 15	Potentiometric titration of amino acids	82
Experiment 16	Estimation of amino acids (the Formol titration)	85
Experiment 17	Isolation of casein	86
Experiment 18	Separation of amino acids of a protein hydrolysate by paper chromatography	87
Experiment 19	Ultra-violet spectra of tyrosine	89
Experiment 20	Quantitative estimation of proteins	90
Experiment 21	Kjeldahl's estimation of nitrogen	91
Experiment 22	Nessler's estimation of nitrogen	94
Experiment 23	Gel filtration of different forms of haemoglobin	95

Chapter 4   Lipids   98

Structure and properties of lipids   98
Separation   102

Experiment 24	Properties of glycerides	102
Experiment 25	Estimation of saponification number	105
Experiment 26	Extraction and separation of neutral lipids	107

Chapter 5   Enzymes   111

The role of enzymes as catalysts   111
Mechanism of enzyme reactions   112
Enzyme kinetics   115
Enzyme assays   118
Enzyme nomenclature   120
Inhibitors   121

Experiment 27	Salivary amylase activity	122
Experiment 28	Succinic dehydrogenase activity	124
Experiment 29	Urease activity. Variation of activity with temperature and pH	126
Experiment 30	Alkaline phosphatase activity	129
Experiment 31	Invertase activity	131

CONTENTS

Experiment 32	Catalase activity	134
Experiment 33	Lipase activity	139

Appendix I pH   141

pH of strong acids and strong bases	142
pH of weak acids and weak bases	142
pH of salt solutions	143
Buffers	145
Indicators	147
Measurement of pH	148

Appendix II Reagents   150

Bibliography   155

Index   157

CHAPTER 1

Introduction

ERRORS

When one measures a physical quantity, one does not expect the value obtained to be exactly equal to the true value. It is, however, important to give some indication of how close one's result is likely to be to the true value.

Accuracy of results is dependent on: (a) the accuracy of the instrument or instruments used, and (b) the precision of the experimenter. Errors in results either are cumulative or can compensate each other.

The error of a sum or difference is equal to the sum of the individual errors:

$$A = B \pm C \quad \text{(sum or difference)}$$
$$A \pm \delta A = B \pm (\delta B) \pm C \pm (\delta C)$$
$$\delta A = \pm [(\delta B) + (\delta C)]$$

The relative error of a product or quotient is equal to the sum of the individual relative errors:

$$A = B \times C \quad \text{(product)}$$
$$A = \frac{B}{C} \quad \text{(quotient)}$$
$$\frac{\delta A}{A} = \pm \left[\frac{(\delta B)}{B} + \frac{(\delta C)}{C} \right]$$

For small errors, it is the general rule that the error in the result can be found by adding the effects of the separate errors.

Random Errors

A random error is one which varies in a set of readings taken from the same experiment; it may be positive or negative.

The magnitude of the random error is dependent on the precision of the experimenter. Random errors can be treated statistically.

Systematic Errors

A systematic error is one which is constant in a set of readings taken from the same experiment. An experiment can be carried out with great precision but the result is inaccurate; the results may all be too high or too low.

Systematic errors are due to inaccurate instruments (the instruments give readings which are too high or too low) or to the design of the experiment. Systematic errors cannot be treated statistically.

Brief Survey of Random Errors in the Measurement of Physical Quantities

The arithmetic mean ($\bar{x}$) of a set of readings of the same quantity is taken as the best value of the quantity:

$$\bar{x} = \frac{\Sigma x}{n}$$

where Σx = sum of individual readings, and n = number of readings.

The error of the result is given by ($\mu - \bar{x}$), where μ is the true value (which is usually unknown). The greater the number of readings (n) one obtains, the closer the arithmetic mean comes to the true value. Therefore the error of the result decreases with the number of readings.

The difference between any individual reading and the arithmetic mean is known as the *deviation from the mean*.

If a large number of readings are determined for the same quantity, the results tend to follow a normal distribution curve, as shown in Figure 1.1.

The dispersion or spread of results about the mean is given in terms of the standard deviation (σ). The standard deviation (σ) = square root of the sum of deviations from the mean squared

INTRODUCTION

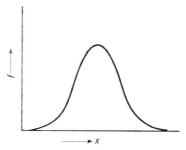

Figure 1.1. A normal curve for the distribution of experimental results; $f = $ frequency of readings of a particular value or range; $n = $ total number of readings; $\Sigma f = n$ (sum of individual frequencies = total number (n))

over the total number of readings:

$$\sigma = \left(\frac{\Sigma d^2}{n}\right)^{\frac{1}{2}} \quad \text{or} \quad \left[\frac{\Sigma(x-\bar{x})^2}{n}\right]^{\frac{1}{2}}$$

where $d = (x - \bar{x}) = $ deviation from the mean.
The smaller the standard deviation with respect to the mean, the smaller the dispersion of results.
The standard error of the mean is given by $\sigma/\sqrt{n}$, i.e. the standard error of the mean decreases by $\sqrt{n}$ as n increases.

CHARACTERISTICS OF THE NORMAL CURVE

1. The mean is denoted by X.
2. The curve is symmetrical about X and has a maximum at X.
3. The curve decreases rapidly to zero as the deviation from the mean becomes large as compared to σ.
4. The points of inflection occur at $x = X \pm \sigma$.
5. The area under the whole curve represents the probability of all possible values and is therefore equal to 1. The area under half the curve (which corresponds to the area under the curve to the left or to the right of the ordinate drawn at X) is equal to 0·5. This represents a probability (P) of 0·5 that a value will be less than X and a P of 0·5 that a value will be greater than X. This is shown in Figure 1.2.

The area under a section of the curve is determined by integrating the normal curve

$$y = \frac{1}{2\pi\sigma^2} \exp\left[-(x-X)^2/\sigma^2\right]$$

between defined limits.

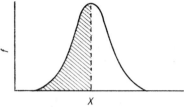

Figure 1.2. A normal curve. The ratio of the shaded area to the total area under the curve shows the probability (P) of a value (x) being less than or equal to X

Consider Figure 1.3, where $x = X+\sigma$.

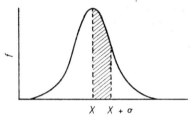

Figure 1.3. A normal curve. The ratio of the shaded area to the total area under the curve shows the probability (P) of a value (x) lying within the range from X to $X+\sigma$

There are tables of ratios of partial areas to the whole area under the normal curve corresponding to various values of X/σ from which the probability of a value falling within or outside a certain range can be determined. It can be seen from these tables that:

The range $\bar{x}\pm\sigma$ covers 68% of the values. (There is a P of 0·68 that a value falls within this range and a P of 0·32 that a value falls outside this range.)

The range $\bar{x}\pm 2\sigma$ covers 95% of the values. (There is a P of 0·05 that a value falls outside this range.)

The range $\bar{x}\pm 3\sigma$ covers 99% of the values. (There is a P of 0·01 that a value falls outside this range.)

MEASUREMENT OF BIOLOGICAL DATA

Biological values, such as the life span of a certain species, values of chemical substances in body fluids, etc., possess no single true value. Normal values for such biological data exist as a range.

In determining normal ranges, it is important that a large sample be taken, so that the sample is representative of the whole population. In large samples the arithmetic mean ($\bar{x}$) of the sample will approach the true mean of the population (μ) and the standard deviation of the sample (s) will approach the standard deviation of the population (σ).

The normal range for biological data is given by $\bar{x} \pm 2s$, where s is the standard deviation of the sample. This range includes 95% of normal values; there is a 5% chance that a normal value falls outside this range.

Significance of Biological Results

A frequent problem is encountered in biological experiments. If a species is placed in a different environment from the normal, it is asked whether this significantly alters the normal values for that species; for example, if a number of rats are subjected to malnutrition, does this significantly alter their growth rate, life expectancy, etc.?

In statistical analysis of this problem the sample taken from the different environment should be large enough so that the mean of the sample approaches the true mean from the different environment.

The mean of the sample from the different environment is considered significantly different from the normal if the difference between the means is equal to or greater than twice the standard error of the mean, i.e.

$$\mu - \bar{x} \geqslant 2 \cdot \frac{\sigma}{\sqrt{n}} \quad \text{or} \quad \frac{\mu - \bar{x}}{\sigma/\sqrt{n}} \geqslant 2$$

where μ = mean of sample from normal environment; $\bar{x}$ = mean of sample from different environment; σ = standard deviation of the mean of sample in normal environment; n = number in sample from different environment; $\mu - x$ = error of the mean; and $\sigma/\sqrt{n}$ = standard error of the mean.

Often σ is unknown and it is assumed that s, the standard deviation of the sample from the different environment, is approximately equal to σ and can be substituted for σ.

The condition

$$\frac{\mu - \bar{x}}{\sigma/\sqrt{n}} \geqslant 2$$

gives a probability (P) of 0·05 or less that the mean ($\bar{x}$) of the sample from the different environment falls outside the normal range, i.e. there is a probability of 0·95 that if the mean does fall outside this range the result is significantly different.

Most workers consider a probability of 0·05 due to chance to be the significant level. Some workers prefer a probability of 0·01 due to chance to be considered the significant level, and this is given by

$$\frac{\mu-\bar{x}}{\sigma/\sqrt{n}} \geqslant 3$$

This method of determining significance of results is more accurate than observing whether the mean of the sample from the different environment falls outside the normal range ($\mu \pm 2\sigma$) because in this method the size of the sample from the different environment (n) is taken into account. The magnitude of n affects the accuracy of the mean from the different environment.

Significant Figures

The accuracy with which one calculates one's results is dependent on the accuracy of the experiment. It is useless to calculate one's results to six significant figures (which gives an error in the region of 0·0005%) if the experiment is only accurate to 1%. For example, a Saponification Number given as 198·137 (when determined titrimetrically) is meaningless. This result presumes an accuracy in the region of 0·0002%. On the other hand, a molarity determined as 0·0922 M should not be approximated to 0·1 M; this introduces an error of about 8%.

In general, experimental results (depending to a certain extent on the accuracy of the experiment) should be given to three significant figures; this gives a maximum error of 0·5%. In this case the example of Saponification Number is reported as 198 and the example of molarity as 0·092 M.

GRAPHS

In experimental sciences graphs are most important in presenting results and in showing the change of one variable with change in another.

In all graphs it is a well-established convention that the independent variable, i.e. the variable whose value is chosen by the experimenter, is plotted along the x-axis ($y = 0$ line) and that the dependent variable, i.e. the variable whose value is then determined, is plotted along the y-axis ($x = 0$ line).

The following points should be noted in plotting graphs.
1. The experimental points should be clearly marked and should not be crowded together.

INTRODUCTION 7

2. It is unnecessary to begin the axes at zero unless the graph represents a straight line (for example, beginning the x-axis scale at 0°C has no physical meaning; the zero on the temperature scale is $-273 \cdot 16°C$ (0 K)). If the graph is a straight line satisfying the relationship $y = mx$, both scales should begin at zero and the line passes through the origin; if, on the other hand, the graph represents a straight line satisfying the relationship $y = mx+c$, only the x-axis scale need begin at zero.
3. It is usual to draw the 'best' smooth curve (or straight line) through the experimental points rather than join each experimental point together. The latter method often leads to a jagged curve; this implies a highly erratic and therefore unlikely relationship between the two variables.
4. The scale should be simple; for example, 1 inch on the graph should represent 1, 2, 5 or 10 units.
5. The scales along the axes should be simple numbers, such as 0·1, 0·2, 0·3, etc., 1, 2, 3, etc. or 10, 20, 30, etc., rather than 100 000, 200 000, 300 000, etc. or 0·0001, 0·0002, 0·0003, etc. The units $\times 10^n$, where n is a positive or negative index, should be clearly marked along the axes.
6. The scales should be chosen so that the gradient of a curve at the initial experimental point or the gradient of a line is in the region of 45°.

Some of these points are displayed in Figure 1.4.

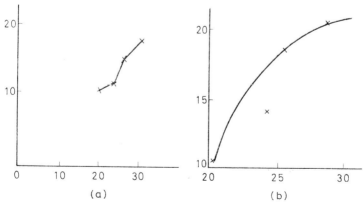

Figure 1.4. (a) displays a poorly drawn graph; (b) displays a well-drawn graph

Another function of graphs is to determine (a) the relationship between the two variables $(y = f(x))$, and (b) the constants which correlate these two variables.

The relationship between the two variables can be reduced basically to two types:

1. The two variables give a straight line relationship or the two variables can be arranged in such a way as to give a straight line relationship: for example, exponential growth and decay.
2. The two variables are related in such a way that they cannot be arranged to give a straight line relationship: for example, the normal distribution curve.

Straight Line Graphs

A straight line is represented by the equation $y = mx+c$ (see Figure 1.5), where m is the gradient and c is the intercept (when $x = 0$, $y = c$). m and c are the constants which correlate the variables x and y.

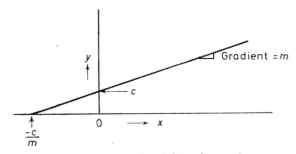

Figure 1.5. A straight line graph satisfying the equation $y = mx+c$

In plotting a graph of this type it is usually only necessary to begin the x-axis scale at zero; if, however, the intercept on the x-axis (which is equal to $-c/m$) is required, it is then necessary to begin the y-axis at zero and to plot the x-axis in the negative as well as the positive direction.

A graph which obeys the equation $y = mx+c$ is the variation of potential with pH in a glass/reference electrode cell (temperature constant) (see Figure 1.6).

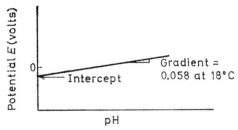

Figure 1.6. An example of a straight line graph satisfying the equation $y = mx+c$

INTRODUCTION

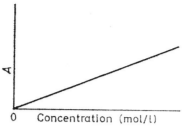

Figure 1.7. An example of a straight line graph satisfying the equation $y = mx$

A straight line is also represented by the equation $y = mx$, where m is the gradient. This line passes through the origin. A graph of this type is the variation of absorbance with the concentration of a substance when the length of the light path is constant (Beer's law). This type of graph is shown in Figure 1.7.

Many non-linear relationships between two variables can be plotted as straight lines. For example:

1. y vs. x^2. (In a relationship of this type, when y is plotted against x, a single loop curve is obtained.) A straight line can be obtained by plotting y against x^2.

When one is plotting a function of x (in this case x^2) along the abscissa, it is preferable that the experimental points be evenly spaced. One should therefore choose values of x which give values of x^2 evenly spaced along the axis. For example, if x^2 is plotted as 0, 10, 20, 30, 40, 50, etc., this roughly corresponds to x having the following values: 0, 3·1, 4·4, 5·4, 6·3, 7·0, etc.

An example of this type of relationship is the initial reaction rate of a bimolecular reaction, which can be expressed as

$$v = kx^2$$

where v = initial rate of reaction, x = concentration of reactants (assumed equal in this case), and k = a constant.

2. A hyperbolic curve can be obtained as a straight line by plotting the reciprocals of the variables.

An example of this is the Lineweaver–Burk plot in enzyme kinetics, where the relationship between v (initial reaction velocity) and $[S]$ (substrate concentration) is given by

$$\frac{1}{v} = \frac{K_M}{V_{max}} \cdot \frac{1}{[S]} + \frac{1}{V_{max}}$$

In this case $1/v$ is plotted against $1/[S]$, the gradient is equal to K_M/V_{max} and the intercept is equal to $1/V_{max}$ (the intercept on the x-axis is equal to $-1/K_M$).

The independent variable should be chosen to give values of $1/[S]$ evenly spaced along the axis. (For example, values of $[S]$ equal to 1, 2, 4, 8, 16, etc., correspond to values of $1/[S]$ of 1, $\frac{1}{2}$, $\frac{1}{4}$, $\frac{1}{8}$, $\frac{1}{16}$, etc.)

3. Exponential growth and decay. Organic growth and decay is proportional to the number of species present at any instant and not to the original number present.

Exponential growth, e.g. population growth, is given by the equation

$$N = N_0 e^{\alpha t}$$

where N_0 = original number at zero time, N = number present at time t, and α = a constant. This function is plotted in Figure 1.8.

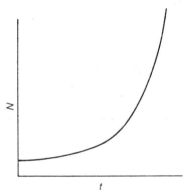

Figure 1.8. An exponential growth curve

Exponential decay, e.g. radioactive decay, is given by the equation

$$N = N_0 e^{-kt}$$

where k = constant and the other symbols have the same meaning as above. This function is plotted in Figure 1.9.

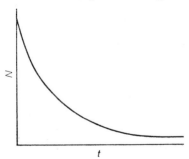

Figure 1.9. An exponential decay curve

INTRODUCTION 11

Exponential curves can be obtained as straight lines by plotting the logarithm of the number at a particular time against time. This relationship is derived as follows:

$$\frac{N}{N_0} = e^{\alpha t} \quad \text{exponential growth}$$

$$\ln \frac{N}{N_0} = \alpha t$$

$$\log_{10} \frac{N}{N_0} = At \quad (A = \text{constant} = 0.4343\alpha)$$

$$\left. \begin{array}{l} \log_{10} N = \log_{10} N_0 + At \\ (\log_{10} N = \log_{10} N_0 - Bt) \quad \text{exponential decay} \\ \phantom{(\log_{10} N = \log_{10} N_0 - Bt)} \quad (B = \text{constant} = 0.4343k) \end{array} \right\} \begin{array}{l} \text{straight line} \\ \text{forms of} \\ \text{exponential} \\ \text{curves;} \\ \text{Figure 1.10} \end{array}$$

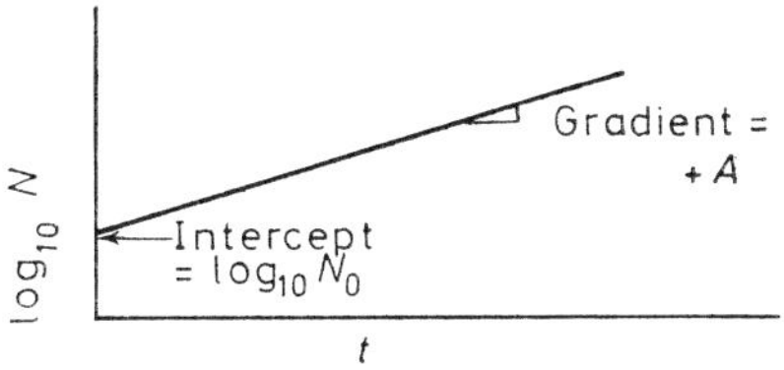

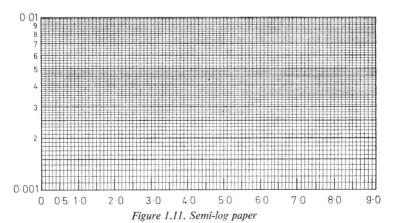

Figure 1.10. *A straight line plot of an exponential curve*

The line has a positive gradient for growth curves, a negative gradient for decay curves and an intercept of $\log_{10} N_0$.

A straight line can also be obtained by plotting N against t on semi-log paper. (Semi-log paper has the y-axis as a log scale and the x-axis as a normal scale; see Figure 1.11.)

Figure 1.11. *Semi-log paper*

4. $y \propto x^n$. In relationships of this type, a straight line can be obtained by plotting log y against log x (log $y = n.\log x$). The line has a gradient of n.

A straight line can also be obtained by plotting y against x on log-log paper. Log-log paper is shown in Figure 1.12.

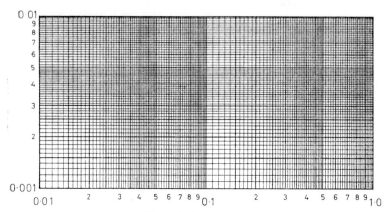

Figure 1.12. Log-log paper

This type of graph is useful in determining the order (n) of a multimolecular reaction.

A straight line is the most accurate graph for determining the constants which correlate two variables.

Non-linear Graphs

Relationships which cannot be converted to a straight line form are of the type

$$y \propto e^{-x^2}$$

Examples of these relationships are:
1. The normal (or gaussian) distribution curve.
2. The elution of a substance from a column. If the elution of a substance from a column does not follow a normal curve, this indicates that the fraction is impure or contains more than one substance.
3. Transmission of light through a monochromator. No monochromator is capable of transmitting light of one particular wavelength only; there is always a normal distribution of transmitted light on either side of the maximum.

CHAPTER 2

Carbohydrates

STRUCTURE OF CARBOHYDRATES

Carbohydrates act as an essential energy source in animal life —for example, starch, glycogen, cane sugar, etc.—and as structural units in animal and plant life—for example, cellulose, chitin, hyaluronic acid, etc.

The word 'carbohydrate' originally signified a group of compounds which contained carbon, hydrogen and oxygen—the latter two elements being present in the same ratio as in water. Many compounds which do not obey this rule now fall into the classification of carbohydrates.

Carbohydrates are divided into three main groups: monosaccharides, oligosaccharides and polysaccharides. Monosaccharides are simple sugars which cannot be hydrolysed into smaller units. They are the building blocks of oligosaccharides and polysaccharides. Oligosaccharides are compound sugars which yield between two and six units of simple sugars on hydrolysis. Polysaccharides are macromolecules which yield a very large number of simple sugars on hydrolysis.

Monosaccharides

Monosaccharides are polyhydroxy compounds characterised by an aldehyde or ketone functional group on the first or second carbon atom, respectively. The simple sugars containing an aldehyde group are called aldoses and those containing a ketone group are called ketoses. Monosaccharides derive their names from the number of carbon atoms in the molecule. Some examples follow.

Trioses	3 carbon atoms
Tetroses	4 carbon atoms
Pentoses	5 carbon atoms
Hexoses	6 carbon atoms

All monosaccharides (except dihydroxy acetone) contain asymmetric carbon atoms; therefore they exist in different optically active forms. Because of the very large number of stereoisomers which occur in carbohydrate chemistry, the simplest monosaccharide which contains an asymmetric carbon atom has been chosen as the reference compound. This compound is glyceraldehyde, an aldotriose, and all other monosaccharides are related to this compound.

```
      H   O                H   O
       \ //                 \ //
        C                    C
        |                    |
   H—C—OH             HO—C—H
        |                    |
      CH₂OH               CH₂OH

   D(+)glyceraldehyde    L(-)glyceraldehyde

      H   O                H   O
       \ //                 \ //
        C                    C
        |                    |
       ,C                   ,C
    H´ |  ▼OH          HO´ |  ▼H
      CH₂OH               CH₂OH
```

The letters D and L denote the absolute configuration on the penultimate carbon atom. The D-series of sugars has the hydroxy group on the penultimate carbon atom written to the right (orientated out of the plane of the paper), whereas the L-series of sugars has the hydroxy group on the penultimate carbon atom written to the left (orientated into the plane of the paper).

N.B. The letters D and L bear no relationship to whether the sugar is dextro- or laevo-rotatory; this is denoted by the symbols (+) and (−).

Most naturally occurring sugars are in the D-series.

Pentoses are important in nucleic acid chemistry—in particular, ribose and 2-deoxyribose. Aldopentoses possess three asymmetric carbon atoms and there are therefore eight stereoisomers: four in the D-series and four in the L-series.

Hexoses are the important sugars in carbohydrate chemistry. The most common naturally occurring hexoses are D-glucose, D-mannose, D-galactose and D-fructose. The first three sugars are aldo-

CARBOHYDRATES 15

```
     H   O              H   O                  CH₂OH
      \\//               \\//                    |
       C                  C                     C=O
       |                  |                     |
   H—C—OH            H—C—OH                HO—C—H
       |                  |                     |
   H—C—OH            HO—C—H                 H—C—OH
       |                  |                     |
   H—C—OH             H—C—OH                H—C—OH
       |                  |                     |
      CH₂OH            H—C—OH                  CH₂OH
                           |
                          CH₂OH

    D-ribose           D-glucose             D-fructose
```

hexoses and the last is a ketohexose. Aldohexoses possess four asymmetric carbon atoms; therefore there are 16 possible stereoisomers. Ketohexoses possess only three asymmetric carbon atoms; therefore there are eight possible stereoforms.

Glucose shows many properties which cannot be explained by the straight chain structure shown above. It displays only some of the reactions characteristic of the aldehyde functional group. For example, it does not form a sodium bisulphite addition compound, an ammonia addition compound or a hemiacetal derivative. Glucose does, however, form an acetal derivative.

A hemiacetal is the compound formed by the addition reaction between an aldehyde group and an alcohol. A hemiketal is the addition compound between a ketone group and an alcohol. An acetal or ketal is the condensation product between a hemiacetal or hemiketal and an alcohol.

```
    H   O                                      OR
     \\//                                       |
      C       +  ROH        ⇌           H—C—OH
      |                                         |
                                          hemiacetal
```

```
     OR                                        OR
      |                                         |
   H—C—OH  +  ROH           ⇌           H—C—OR   +  H₂O
      |                                         |
                                           acetal
```

The hemiacetal and hemiketal formation is a readily reversible reaction and a hemiacetal or hemiketal is often referred to as a potential aldehyde or ketone. Acetals and ketals are hydrolysed back to carbonyl compounds with dilute mineral acids.

Tanret (1895) showed that glucose displayed three different specific rotation values in solution. When glucose is dissolved in water, it initially shows a specific rotation value of $+113°$, which gradually decreases to $+52.5°$, whereas glucose recrystallised from pyridine or concentrated acetic acid initially shows a specific rotation of $+19°$, which gradually increases to $+52.5°$. This phenomenon of mutarotation indicates that glucose contains another asymmetric carbon atom which is not denoted by the straight chain formula.

Emil Fischer interpreted these data by stating that glucose exists as a ring structure in the form of an internal hemiacetal between the carbonyl group on the first carbon atom and the hydroxy group of the fifth carbon atom.

```
     H   OH                    HO    H
      \ /                        \ /
       C————————┐                 C————————┐
       |        |                 |        |
   H—C—OH       |             H—C—OH       |
       |        |                 |        |
   HO—C—H       O             HO—C—H       O
       |        |                 |        |
   H—C—OH       |             H—C—OH       |
       |        |                 |        |
   H—C——————————┘             H—C——————————┘
       |                         |
     CH₂OH                     CH₂OH

   α-D-glucose                β-D-glucose
```

The above structure gives glucose another asymmetric carbon atom on the first carbon atom. Glucose exists in solution as an equilibrium mixture of the α- and β-forms with traces of the free aldehyde form. This unidentified orientation is indicated by a wavy line in the structures that follow.

In the Fischer notation of sugars, it appears that the C—O bond is longer than the C—C bond. This in fact is not the case, as is verified by X-ray data.

W. N. Haworth (1929) proposed ring structures of sugars based on the heterocyclic compounds pyran and furan.

```
      5                         4    O    1
       \—O—\                     \  / \  /
       /    \                     \/   \/
      4      1                    /\   /\
       \    /                    /  \ /  \
        \  /                    3    2
       3    2

       pyran                     furan
   A 6-membered ring         A 5-membered ring
```

Glucose and all aldohexoses exist predominantly in the pyranose form with traces of the furanose form. The pyranose form is

CARBOHYDRATES 17

α-D-glucopyranose　　Straight chain aldehyde form　　β-D-glucopyranose

the hemiacetal fomred between the C=O of C_1 and the —OH group of C_5. Ketohexoses and aldopentoses exist in the furanose form, the internal hemiketal or hemiacetal formed between the carbonyl group of C_2 and C_1, respectively, and the hydroxy group of the penultimate carbon atom.

Oligosaccharides

Oligosaccharides consist of monosaccharide units condensed together through an acetal or glycosidic linkage. The most important oligosaccharides are all disaccharides: maltose, lactose, sucrose and cellobiose. Their structural formulas are indicated below:

maltose
4-glucose-1-α-glucoside

lactose
4-glucose-1-β-galactoside

sucrose
1-α-glucosyl-2-β-fructoside

cellobiose
4-glucose-1-β-glucoside

Maltose, lactose and cellobiose all possess a potential free aldehyde group, therefore qualifying as *reducing* sugars. Moreover, they display many characteristics of the aldehyde functional group. Sucrose, on the other hand, is a *non-reducing* sugar.

Maltose and cellobiose are breakdown products of starch and cellulose, respectively. Cellobiose is indigestible in humans because the body does not possess enzymes for hydrolysing the β-glucoside bond.

Polysaccharides

Starch is the reserve source of carbohydrates in plants. It is a polymer of glucose and varies in molecular weight between 10 000 and 1 000 000.

Starch consists of two constituents: amylose and amylopectin.

Amylose, molecular weight 10 000–20 000, usually constitutes the *minor* component of starch. It is a straight chain polymer of glucose; the units are linked α-1-4, e.g.

or as represented in Figure 2.1. Amylose is soluble in hot water and gives a blue colour with iodine. It is a non-reducing carbohydrate.

CARBOHYDRATES

Figure 2.1. Amylose, a straight chain polymer of glucose

Amylopectin, molecular weight 500 000–1 000 000, usually constitutes the *major* component of starch. It is a branched chain polymer of glucose. The straight chain portion of the polymer consists of glucose units linked α-1-4, as in amylose, and positions of branching occur at 1-6-linkages:

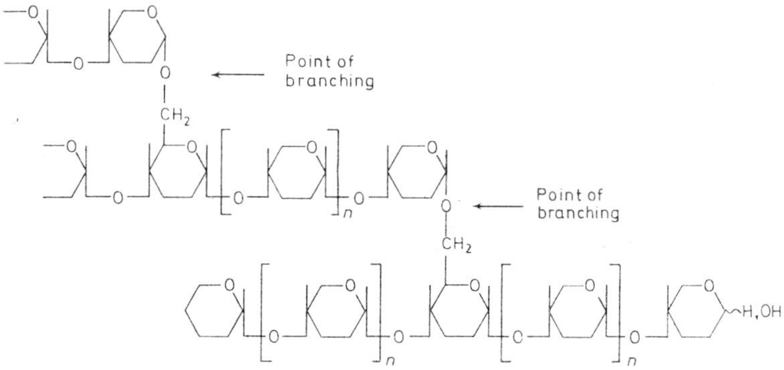

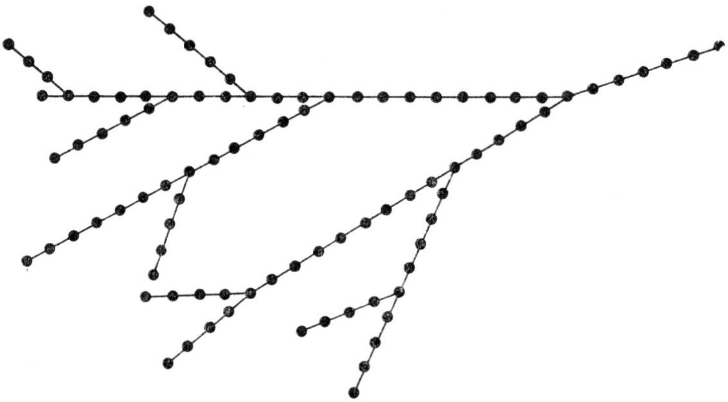

Figure 2.2. Amylopectin, a branched chain polymer of glucose

or as represented in Figure 2.2. Amylopectin is insoluble in hot water and gives a reddish-black colour with iodine.

Glycogen is the reserve source of carbohydrate in most mammals. It is also a polymer of glucose and varies in molecular weight between 1 000 000 and 5 000 000. Its structure is very similar to that of amylopectin, except that the degree of branching is much greater.

Dextrins are compounds formed by the partial hydrolysis of starch. These compounds can be repolymerised by various bacteria or fungi to form dextrans, an example of which is Sephadex (See p. 69). Dextrans have found important uses as gel filtration media and in the determination of molecular weights of macromolecules.

REACTIONS OF SUGARS

1. Dehydration with Mineral Acids

On being heated with concentrated sulphuric acid, sugars are completely dehydrated to form carbon.

Under milder conditions, sugars are partially dehydrated, with the elimination of three molecules of water, to form furfural or a derivative of furfural; e.g.

Pentoses $\xrightarrow{H^+}$ furfural

Hexoses $\xrightarrow{H^+}$ 5-hydroxy methyl furfural

Furfural and its derivatives condense with phenols to form brightly coloured compounds. This characteristic is the basis of many qualitative tests (see Table 2.1).

Table 2.1. REACTION OF SUGARS WITH PHENOLS

Test	Concentrated mineral acid	Phenol	Positive reaction	Colour of compound
Molisch's	H_2SO_4	α-naphthol	all carbohydrates	purple
Seliwanoff's	HCl	resorcinol	ketoses and sucrose	red

Table 2.1 (cont.)

Test	Concenrated mineral acid	Phenol	Positive reaction	Colour of compound
Bial's	HCl	orcinol	pentoses and uronic acids	green
Tollen's	HCl	α-naphtho-resorcinol	uronic acids	blue

N.B. Uronic acids are sugars in which the —CH_2OH group of the ultimate carbon atom has been oxidised to a —CO_2H group.

2. Action of Alkali

Monosaccharides undergo rearrangement around the first and second carbon atom with dilute alkalis. The rearrangement proceeds through an enol intermediate:

D-glucose ⇌ [enol intermediate] ⇌ D-fructose / D-mannose

Lobry de Bruyn rearrangement

Oligosaccharides and polysaccharides are not hydrolysed by dilute alkalis, as the OH⁻ ions would destroy the structure of the constituent simple sugars.

Strong caustic alkalis cause polymerisation of monosaccharides.

3. Oxidation

The carbonyl group of sugars is readily oxidised, by mild oxidising agents such as cupric ions (Cu^{2+}), silver ions (Ag^+), etc., to yield a variety of products. Monosaccharides and oligosaccharides with an uncombined potential aldehyde or ketone group are therefore reducing agents and are referred to as reducing sugars.

The reduction of cupric ions in solution is the basis of Fehling's, Benedict's and Barfoed's tests. These reagents vary in the pH of the medium: Fehling's reagent consists of Cu^{2+} ions in a strongly alkaline medium, Benedict's reagent contains Cu^{2+} ions in a mildly alkaline medium; and Barfoed's reagent comprises Cu^{2+} ions in a mildly acid medium. Benedict's test is more specific for reducing sugars than Fehling's. Barfoed's test is specific for reducing monosaccharides.

Bromine water and iodine solution in an alkaline medium oxidise aldoses to aldonic acids. (Aldonic acids are compounds in which the —CHO group of the aldose has been oxidised to a —CO_2H group.)

50% nitric acid oxidises aldoses to a single dicarboxylic acid and ketoses to two carboxylic acids.

Periodate ions (IO_4^-) oxidise sugars and cause bond cleavage between adjacent carbon atoms. This type of oxidation is discussed in Experiment 8.

4. Osazone Formation

All α-hydroxy carbonyl compounds react with phenylhydrazine to form osazones. The initial reaction is the condensation of the carbonyl group with phenylhydrazine followed by the oxidation of the α-hydroxy group to a carbonyl group, which subsequently condenses with a further molecule of phenylhydrazine.

All reducing sugars form osazones with phenylhydrazine. Osazone formation destroys the configuration on the first and second carbon atoms. Therefore sugars which vary only in their configuration about these two carbon atoms form the same osazone; for example, glucose, mannose and fructose form the same osazone.

Osazones have characteristic crystalline forms which therefore afford a method of identifying the sugars from which the osazone is derived. In addition, osazones of different sugars form at different rates and therefore observation of the time taken for crystals to form also gives a valuable indication of the parent sugar, especially if the sugars form the same osazone.

$$\underset{\text{D-glucose}}{\overset{H}{\underset{R}{\overset{|}{\underset{|}{H-C-OH}}}}\overset{\diagup O}{\diagdown}} + \phi NH.NH_2 \longrightarrow \underset{\text{a phenyl hydrazone}}{\overset{H}{\underset{R}{\overset{|}{\underset{|}{H-C-OH}}}}\overset{\diagdown}{\diagdown} C=N.NH.\phi} + H_2O \xrightarrow{\phi NH.NH_2}$$

$$\overset{H}{\underset{R}{\overset{|}{\underset{|}{C=O}}}}\overset{\diagdown}{\diagdown} C=N.NH.\phi + \phi NH_2 + NH_3 \xrightarrow{\phi NH.NH_2} \underset{\text{an osazone}}{\overset{H}{\underset{R}{\overset{|}{\underset{|}{C=N.NH\phi}}}}\overset{\diagdown}{\diagdown} C=N.NH.\phi}$$

5. Reactions of Hydroxy Groups

The hydroxy groups in sugars can be phosphorylated, methylated, acetylated, benzoylated, etc.

EXPERIMENT 1
QUALITATIVE TESTS FOR CARBOHYDRATES

Reagents

  0·5% (w/v) aqueous sugar solutions (listed overleaf)
* Molisch's reagent
* Benedict's qualitative reagent
* Barfoed's reagent
* Seliwanoff's reagent
* Foulger's reagent
* Bial's reagent
  2 M $NaNO_3$
  Dil. iodine solution
  2 M NH_4OH
  5% w/v aqueous methylamine hydrochloride
  20% w/v NaOH
  Conc. HCl
  Phenylhydrazine hydrochloride
  Sodium acetate
  Glacial acetic acid
* Glucose oxidase reagent

Procedure

Apply the qualitative tests listed below to 0·5% aqueous solutions of the following substances: starch, dextrin, glucose, fructose, mannose, xylose, maltose, sucrose and lactose.

Molisch's test	General test for all carbohydrates, whether pure or associated with protein
Iodine test	Test for macromolecules
Benedict's test Silver mirror test	Test for all reducing sugars
Barfoed's test Fearon's test	Distinguishing tests between reducing monosaccharides and disaccharides
Seliwanoff's test Foulger's test	Tests for keto-sugars
Osazone formation	Reaction of reducing sugars
Glucose oxidase test	Test for glucose

MOLISCH'S TEST

Place 2 ml of each carbohydrate solution in a test-tube and add a few drops of Molisch's reagent and mix well. Then add 1–2 ml of conc. H_2SO_4 *carefully* down the side of the tube so as to form a layer beneath the aqueous solution. A purple ring at the interface indicates a positive reaction.

IODINE TEST

Place 2 ml of each carbohydrate solution in a test-tube and add 3 drops of dilute iodine solution. Unbranched macromolecules give a blue colour and branched macromolecules a reddish-black colour.

BENEDICT'S TEST

To 5 ml of Benedict's solution (cupric sulphate in a mild alkaline medium) add about 1 ml of each carbohydrate solution. Boil in a water-bath for 2 min. The production of a green, yellow or red precipitate indicates a positive reaction for reducing sugars.

SILVER MIRROR TEST

To about 1 ml of silver nitrate solution in a series of grease-free test-tubes (previously rinsed with acetone) add dilute ammonium hydroxide until the precipitate formed just redissolves. Add 1 ml of carbohydrate solution to each tube, and warm in a water-bath at 70°C for 5 min. A silver mirror is formed with reducing sugars.

N.B. Care must be taken in the use of ammoniacal silver nitrate, as the reagent is explosive. Do not store this reagent.

BARFOED'S TEST

To 5 ml of Barfoed's reagent (cupric sulphate in a weak acid medium) add 1 ml of carbohydrate solution. Boil in a boiling water-bath for 3–4 min. A red precipitate or definite red cloudiness is a positive reaction.

FEARON'S TEST

Place 4 ml of each carbohydrate solution in a test-tube and add 3–4 drops of 5% aqueous methylamine hydrochloride. Boil in a water-bath for 30 s. Cool, and add 4–5 drops of 20% NaOH. A red coloration is positive for reducing disaccharides.

N.B. Only perform the above two tests on sugars which have given a positive reaction with Benedict's reagent.

SELIWANOFF'S TEST

To 5 ml of Seliwanoff's reagent add 5–6 drops of carbohydrate solution. Boil in a water-bath for 30 s. A red coloration or red precipitate is positive for keto-sugars and for disaccharides containing a keto-sugar.

FOULGER'S TEST

To 2 ml of sugar solution add 1 ml of conc. HCl and 1 ml of Foulger's reagent (stannous chloride, urea and dilute sulphuric acid). Boil for 2 min. A blue coloration is positive for keto-sugars.

BIAL'S TEST

To 3 ml of Bial's reagent add 1 ml of sugar solution. Boil for 1 min. A green coloration is positive for pentoses and uronic acids.

OSAZONE FORMATION

Mix together, in a series of test-tubes, one spatula-full (about 0·5 g) of phenylhydrazine hydrochloride, 2 spatulas (about 1 g) of sodium acetate, 2 drops of glacial acetic acid and 2–3 ml of sugar solution. Warm the tubes in a boiling water-bath for 1 min in order to dissolve the solids, and, if the solution is not clear, filter into a clean test-tube. Heat the mixtures in a boiling water-bath for 30 min. During this period, examine the tubes at 5-min intervals and note in which tubes, and at what times, formation of crystals occurs. If crystallisation has begun within 30 min, remove the tubes from the water-bath. If crystallisation has not begun after 30 min, allow the tubes to cool slowly in the water-bath, and crystallisation may be hastened by scratching the sides of the tube with a glass rod. If globules of oil are present, filter the mixture into a clean test-tube.

Examine the crystalline osazones under the microscope, sketch the form of the crystals and indicate to what extent the different sugars may be identified by this method.

GLUCOSE OXIDASE TEST

Glucose oxidase reagent contains glucose oxidase, peroxidase and o-tolidine. The latter reagent is an oxygen acceptor and changes from a colourless reduced form to a blue oxidised form. In this test, glucose is oxidised by glucose oxidase to gluconic acid. Hydrogen peroxide is also formed in the reaction and it is decomposed by peroxidase to water and oxygen. The liberated oxygen is accepted by o-tolidine and therefore a blue coloration indicates a positive reaction. Because of the specificity of the enzyme, this test is positive only for glucose. It is therefore a useful test for distinguishing between glucose and other aldohexoses and between glucose and fructose.

Place 1 ml of each carbohydrate solution in a test-tube and add 3 ml of glucose oxidase reagent. Mix gently for 10 s and then leave for 5 min. Note any colour change.

Tabulate your results.

Table 2.2 QUALITATIVE SCHEME FOR CARBOHYDRATE CHEMISTRY

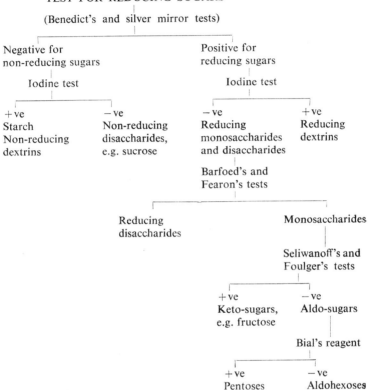

EXPERIMENT 2
ESTIMATION OF GLUCOSE BY IODINE OXIDATION IN ALKALINE CONDITIONS

$C_5H_{11}O_5.CHO + I_2 + 3\,NaOH = C_5H_{11}O_5.CO_2Na + 2\,NaI + 2\,H_2O$
Glucose   Sodium gluconate

Glucose is oxidised to gluconic acid by alkaline iodine solution.

Excess iodine and alkali are added to the glucose solution and the excess iodine is then back-titrated with standardised sodium thiosulphate. Care must be taken not to add too large an excess of the iodine, as this will cause further oxidation of the sodium gluconate formed (about 2 ml of 0·05 M I_2 is a suitable amount).

Reagents

Standardised sodium thiosulphate solution (approx. 0·1 M)
Iodine solution (approx. 0·05 M)
Sodium hydroxide solution (approx. 0·05 M)
2 M HCl
Methyl red
Unknown glucose solution (4–10 g/l)
1% w/v soluble starch

Procedure

Standardise the iodine solution with the standard sodium thiosulphate solution.

Next pipette, in triplicate, 25 ml of the 'unknown strength' glucose solution into a conical flask. Add iodine solution from a burette (about 5 ml at a time), shaking the flask well and leaving it to stand for at least 1 min between each addition. At each addition of iodine, add aliquots of the given sodium hydroxide from a measuring cylinder equivalent to three times the volume of iodine added. Continue the addition of I_2/NaOH until a suitable excess has been added. (An excess of I_2/NaOH can be noted by the appearance of a pale yellow solution due to the formation of sodium hypoiodite.) Record accurately the volume of iodine solution added. Add 2 M HCl to the flask until the mixture is just acid to methyl red (about pH 5). Place the standard sodium thiosulphate in a microburette and titrate the excess iodine against the thiosulphate. 1% soluble starch may be used as an indicator near the end-point.

Calculation

$$\text{Molarity of } Na_2S_2O_3 \text{ solution} = M_{Na_2S_2O_3}$$
$$\text{Molarity of } I_2 \text{ solution} = M_{I_2}$$

From the equation

$$2\ Na_2S_2O_3 + I_2 = 2\ NaI + Na_2S_4O_6$$
$$1 \text{ mol } I_2 \equiv 2 \text{ mol } Na_2S_2O_3$$

$$\text{Volume of excess } I_2 = \frac{1}{2} \times \frac{(\text{titre}_{Na_2S_2O_3}) \times M_{Na_2S_2O_3}}{M_{I_2}}$$

Amount of I_2 required to oxidise 25 ml glucose solution

$= $ (total vol. $I_2 - $ excess vol. $I_2) \times M_{I_2}$

$= $ mmol $I_2/25$ ml glucose solution

$= $ mmol glucose/25 ml glucose solution (1 mol $I_2 \equiv$ 1 mol glucose)

Concn. of glucose soln., in g/l

$= $ (total vol. $I_2 - $ excess vol. $I_2) \times M_{I_2} \times \dfrac{1000}{25} \times \dfrac{180}{1000}$

EXPERIMENT 3
SUGAR CONTENT OF FRUIT

The sweetness of fruit is mainly due to 'invert sugar': an equimolar mixture of fructose and glucose. Some fruits also contain traces of sucrose.

In this experiment sugar present in oranges is extracted with water, and the glucose content of 'invert sugar' is determined titrimetrically by oxidation with iodine in alkaline conditions. An aliquot of the sugar extract is then hydrolysed with acid and the glucose content of hydrolysed sucrose is estimated as before. (The protein content of oranges is very small and therefore it is unnecessary to prepare protein-free samples before estimating the sugar.)

N.B. This method estimates all sugars with an aldehyde or potential aldehyde group but glucose is the only aldose present in any substantial amount in oranges.

Reagents and Apparatus

Oranges
Standardised iodine solution (approx. 0·05 M)
Standardised sodium thiosulphate solution (approx. 0·1 M)
0·05 M NaOH
1% w/v soluble starch
2 M HCl
2 M NaOH
Methyl red
Homogeniser

Procedure

Peel an orange, and remove the skin and pips from the sections. Weigh 50 g of skinned and de-pipped orange sections. Homogenis- orange sections in 50 ml of water and transfer homogenate quane titatively to two 50-ml centrifuge tubes. Leave the tubes for about 10 min, stirring occasionally with a glass rod. Centrifuge for 3 min. Decant supernatants quantitatively to a 100-ml volumetric flask and make up to the mark.

Pipette, in triplicate, 5 ml of sugar extract into a conical flask. Determine glucose content by titration with I_2/NaOH as in Experiment 2.

Pipette 25 ml of sugar extract into a 250-ml conical flask, add 5 ml of 2 M HCl and heat in boiling water-bath for 20 min. Cool, adjust pH to about 6 with 2 M NaOH, using universal indicator paper. Transfer hydrolysate to a 50-ml volumetric flask and make up to the mark. Pipette, in triplicate, 5 ml of hydrolysed sugar extract into a conical flask and determine the glucose content by titration with I_2/NaOH as before.

Calculation

1. Glucose content in 5 ml sugar extract (2·5 g fruit)
   = (total vol. I_2 − excess vol. I_2) × M_{I_2} × 180 mg glucose
   Calculate glucose content in 100 g fruit
   Invert sugar concn. = 2 × glucose concn.
   Express result as g invert sugar/100 g fruit
2. Total glucose content in 5 ml hydrolysed sugar extract (1·25 g fruit)
   = glucose content from invert sugar + glucose content from sucrose
   Calculate g invert sugar derived from sucrose/100 g fruit
   342 g sucrose gives 360 g invert sugar
   Invert sugar × 0·95 = sucrose
   Express sucrose concn. as g sucrose/100 g fruit

EXPERIMENT 4

ESTIMATION OF SUGARS BY BENEDICT'S METHOD

Benedict's quantitative reagent is a modification of the qualitative reagent. It contains the same constituents, namely cupric sulphate, sodium carbonate and sodium citrate. It also contains potassium thiocyanate and a small amount of potassium ferrocyanide. The inclusion of the citrate prevents the precipitation of cupric carbonate by chelating the Cu^{2+} ions. The thiocyanate causes the precipitation of white cuprous thiocyanate rather than red cuprous oxide on reduction of cupric ions, which enables the end-point of the titration, the transition from blue to white, to be readily observable. Methylene blue may be used as an additional indicator. The small amount of potassium ferrocyanide prevents any reoxidation of cuprous ions.

The reduction of Cu^{2+} ions by sugars is a non-stoichiometric reaction and is only constant over a small range of sugar concentration. For this reason, to obtain accurate results, the volume of sugar added must lie within 6–12 ml/10 ml Benedict's reagent. If the preliminary titre falls outside this range, the sugar solution must be appropriately diluted or concentrated before further titrations are performed.

N.B. A non-stoichiometric reaction is a reaction which does not follow a defined pathway and cannot be described by an equation either qualitatively or quantitatively.

Reagents

* Benedict's quantitative reagent
  Named sugar of unknown concentration
  Sodium carbonate
  Methylene blue

Procedure

Fill the burette with the sugar solution. Pipette, using a safety filler, 10 ml of Benedict's quantitative reagent into a 100-ml conical flask and add 2 g of anhydrous sodium carbonate and two pieces of porous pot; shake the mixture well to suspend the sodium carbonate. Clamp the flask at an angle beneath the burette and gently heat the mixture to boiling. Keep the contents of the flask boiling steadily throughout the experiment. If the volume of the flask is reduced through evaporation, add distilled water to keep the volume constant.

Carry out a preliminary titration by adding 3-ml aliquots of sugar solution to the flask and allow the mixture to boil for 30 s–1 min between each addition. (Disaccharides are oxidised more slowly than monosaccharides and should therefore be allowed to stand for 1 min between each addition; 30 s is adequate for monosaccharides.) If the blue colour is completely discharged on addition of 6 ml of sugar solution, dilute the sugar twofold before carrying out another preliminary titration. If, on the other hand, the colour is not discharged on addition of 12 ml of sugar solution, concentrate the solution twofold before carrying out a further preliminary titration.

When the preliminary titre lies between 6 and 12 ml, carry out, in triplicate, an accurate titration. Methylene blue may be used as an additional indicator near the end-point.

N.B. Non-reducing sugars, e.g. sucrose, are estimated by first hydrolysing the sugar with dilute HCl to its component monosaccharides and then titrating the hydrolysed mixture with Benedict's reagent.

Calculation

$$10 \text{ ml Benedict's} \equiv 0{\cdot}0200 \text{ g glucose}$$
$$0{\cdot}0202 \text{ g fructose}$$
$$0{\cdot}0271 \text{ g lactose}$$
$$0{\cdot}0282 \text{ g maltose}$$
$$0{\cdot}0196 \text{ g sucrose}$$
$$10 \text{ ml Benedict's} \equiv x \text{ g sugar} \equiv y \text{ ml sugar solution}$$

Express the results in g/l

EXPERIMENT 5
PHOTOMETRY: COLORIMETRIC ESTIMATION OF SUGARS

Photometry

Many assays in biochemistry depend on measuring light absorption of a substance in solution in the visible and ultraviolet region of the electromagnetic spectrum. The estimation is based on comparison of the light absorbed by a substance of unknown concentration with the light absorbed by the same substance of known concentration. This method can be used in the estimation of coloured substances; those which absorb only in the UV, such as proteins and nucleotides, viz. substances containing an aromatic ring structure; and colourless substances such as sugars, which can undergo a series of chemical reactions to give a coloured derivative. The advantage of this type of assay is that it is very sensitive to micro quantities of solute.

When white light is passed through a coloured solution, certain wavelengths of light are selectively absorbed and the resultant colour is due to the light which is transmitted. Maximum absorption of coloured solutions occurs in the region of opposite colour, i.e. red solutions absorb maximum light in the green region, etc. (Figure 2.3).

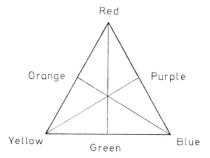

Figure 2.3. Diagram showing the relationship between the colour of a solution (e.g. red) and the colour of the light absorbed (e.g. green)

The amount of light absorbed by a solution is governed by the factors of path length through which the light travels and the concentration of the solution. These properties are formalised in two laws, namely Lambert's law and Beer's law.

Lambert's law

Lambert's law states that the same proportion of incident light is absorbed per unit thickness irrespective of its intensity, and that each successive unit layer absorbs the same proportion of light falling on it. For example, if the incident light is 100%, and 50% of the incident light is absorbed per unit layer, then the intensity of light will decrease as follows: 50%, 25%, 12·5%, 6·25%, 3·13%, etc. (see Figure 2.4).

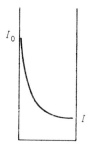

Figure 2.4. Exponential decrease of incident through a medium: I_0, incident light; I, emergent light

Therefore the light decreases exponentially through the cell and this is expressed mathematically as

$$I = I_0 e^{-kl}$$

where l = path length, and k = constant;

$$\frac{I}{I_0} = e^{-kl} \quad \left(\frac{I}{I_0} = \text{transmittance, } T\right)$$

Taking logs to the base 10 on each side of the equation, we obtain

$$\log_{10} \frac{I_0}{I} = kl \cdot \log_{10} e$$

$$\log_{10} \frac{I_0}{I} = Kl$$

where K = constant = $0.4343k$. $\log_{10} \frac{I_0}{I}$ is called either absorbance (A) or optical density (O.D.) A is directly proportional to l (path length).

Beer's law

This law states that the absorption of light is directly proportional to the number of absorbing molecules (the transmittance decreases exponentially with the number of absorbing molecules):

$$\log_{10} \frac{I_0}{I} \propto c$$

where c = concentration in mol/l.

Combining these two laws, we obtain:

$$\log_{10} \frac{I_0}{I} \propto cl$$

$$\log_{10} \frac{I_0}{I} = \varepsilon cl$$

where ε = molar extinction coefficient (1 mol^{-1} cm^{-1}). It is numerically equal to the absorbance of a molar solution with a path length of 1 cm.

Provided that the path length is kept constant, the absorbance of a solution is directly proportional to its molar concentration, if Beer's law is obeyed.

Lambert's law holds for all cases, but Beer's law is usually only obeyed for dilute solutions. At certain concentrations association of the absorbing molecules is thought to occur, which causes a tailing off in the absorption of light; this is shown in Figure 2.5.

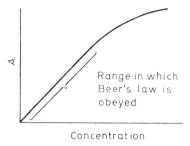

Figure 2.5. Variation of absorbance with concentration. Estimations should be carried out in the concentration range in which Beer's law is obeyed

It is usual to plot a standard curve (A against c) to determine the concentration range in which Beer's law is obeyed, as this varies for different substances.

Within the range in which Beer's law is obeyed and provided that the path length is kept constant, the concentration of a solution can be determined by comparison of its light absorption with that of a solution of the same substance of known concentration:

$$A_{unknown} \propto c_{unknown}$$

$$A_{standard} \propto c_{standard}$$

$$\frac{A_{unknown}}{A_{standard}} = \frac{c_{unknown}}{c_{standard}}$$

$$c_{unknown} = \frac{A_{unknown}}{A_{standard}} \times c_{standard}$$

Since absorbance is a logarithmic scale to the base 10, an increase of 1 absorbance unit represents a tenfold decrease in light transmission. For example:

when $A = 0$, 100% light is transmitted
when $A = 1$, 10% light is transmitted
when $A = 2$, 1% light is transmitted

For absorbance values greater than 1, when less than 10% of the light is transmitted, a slight error in the transmitted light causes a large error in the absorbance value. Slight errors in transmitted light can be caused by a number of factors: for example, stray light in the instrument, light reflected by the absorbing molecules and by the surfaces of the instrument, fluorescence, etc.

The most accurate range in photometry is 20–85% transmission (A in the range 0·1–0·8).

THE INSTRUMENT

A colorimeter is an instrument for measuring light absorption. It has a coloured filter for producing monochromatic light. This produces a broad bandwidth of light. A spectrophotometer is an instrument for measuring light absorption or transmission, which has an accurate means of producing monochromatic light, such as a prism or a diffraction grating. This produces a narrow bandwidth of light. The relative bandwidths of emergent light in a colorimeter and a spectrophotometer are shown in Figure 2.6.

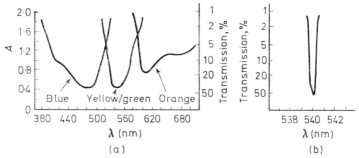

Figure 2.6. Bandwidth of emergent light through (a) a coloured filter and (b) a monochromator

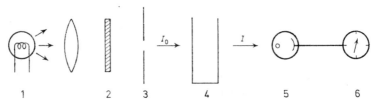

Figure 2.7. A colorimeter: 1, light source (tungsten lamp); 2, monochromator coloured filter; 3, slit; 4, optical cell; 5, photo-electric cell; 6, galvanometer

The essential features of a colorimeter are represented in Figure 2.7.

A spectrophotometer is also a useful tool in qualitative chemistry and biochemistry. The determination of the absorption spectrum (absorbance against λ) of a compound gives a pattern which is characteristic of a compound or a class of compounds, and can be used in their identification.

Nelson's Colorimetric Determination of Sugars

Cupric ions are reduced by sugars to cuprous ions, which quantitatively reduce phosphomolybdic acid to a blue complex.

REAGENTS AND APPARATUS

* Alkaline copper reagent
* Phosphomolybdic acid reagent
  Standard glucose solution containing 100 µg/ml
  Glucose solution of 'unknown strength' (30–70 µg/ml)
  Colorimeter or spectrophotometer

PROCEDURE

Pipette, in duplicate, 0·25, 0·5, 0·75, 1·0, 1·5 and 2·0 ml of standard glucose solution into a series of test-tubes. Make the volumes up to 2 ml with water. Also prepare a blank containing 2 ml of water. In addition, prepare two tubes containing 2 ml of 'unknown strength' glucose solution. Add 2 ml of alkaline copper reagent to each tube, mix and plug each tube with a small piece of cotton wool—this prevents reoxidation of Cu^+ ions. Heat the tubes in a boiling water-bath for exactly 8 min, then allow the tubes to cool in cold water. Add 2 ml of phosphomolybdic reagent to each tube and shake well. After about 2 min, dilute each tube to 10 ml with water. Read absorbance against the reagent blank at 630 nm (red filter).

Plot a standard curve for glucose and comment on the concentration range in which Beer's law is obeyed. Read the concentration of the unknown from the standard curve and express the concentration of the unknown in g/l.

EXPERIMENT 6
ESTIMATION OF LACTOSE IN MILK

Lactose is the major carbohydrate in milk. It constitutes 4–6% of cow's milk and 2–3% of human milk. It is a reducing sugar, as it possesses an uncombined potential aldehyde group on the glucose unit. Therefore lactose concentration can be determined by any of the methods suitable for reducing sugars.

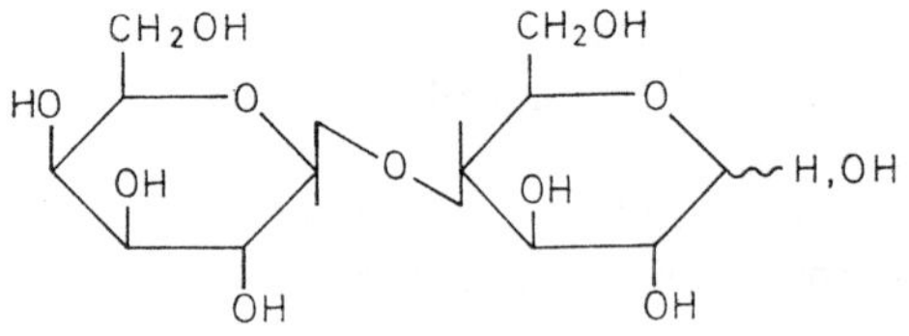

Lactose or 4-glucose-1-β-galactoside

In this method protein-free milk is prepared by the Folin–Wu method; then the lactose concentration is determined by Nelson's colorimetric method.

Reagents and Apparatus

* Fat-free milk
10% w/v sodium tungstate
1/3 M H_2SO_4
* Alkaline copper reagent
* Phosphomolybdic acid reagent
Lactose stock solution containing 1 g lactose in 100 ml 0·2% w/v benzoic acid
0·2% w/v benzoic acid
Folin–Wu tubes
Colorimeter or spectrophotometer

Procedure

To 1 ml of fat-free milk in a 100-ml volumetric flask, add 2 ml of 10% sodium tungstate solution, then add, slowly and with constant shaking, 2 ml of 1/3 M H_2SO_4. Make the mixture up to the mark and allow it to stand for 5 min. Filter the mixture through a Whatman No. 42 filter paper.
Into Folin–Wu tubes, pipette, in duplicate:

1 ml filtrate plus 1 ml distilled water
2 ml lactose standard containing 0·6 mg lactose
2 ml distilled water as a reagent blank

Prepare the lactose standard from the lactose stock solution containing 1 g/100 ml by pipetting 3 ml into a 100-ml volumetric flask and making up to the mark with 0·2% benzoic acid.
To each tube add 2 ml of alkaline copper reagent and heat in a boiling water-bath for 8 min. Cool and, with thorough mixing, add

4 ml of phosphomolybdic acid reagent. Allow the tube to stand for 1 min, then dilute the mixtures to the 25-ml mark with diluted 1 : 4 phosphomolybdic acid. Read absorbance against reagent blank at 630 nm (red filter).

Calculation

$$\frac{A_{unknown}}{A_{standard}} = \frac{c_{unknown}}{c_{standard}}$$

$c_{unknown}$ (lactose concn. in 0·01 ml milk)

$$= \frac{A_{unknown}}{A_{standard}} \times \frac{0 \cdot 6}{1000} \text{ g lactose}$$

$$\text{Lactose concn. g}/100 \text{ ml} = \frac{A_{unknown}}{A_{standard}} \times \frac{0 \cdot 6}{1000} \times \frac{100}{0 \cdot 01}$$

EXPERIMENT 7
POLARIMETRY

Compounds which possess one or more asymmetric carbon atoms have the power of rotating the plane of polarised light either in solution or in the crystalline state. Such compounds are said to be optically active. The measurement of this property is called polarimetry.

The magnitude and direction in which the plane of polarised light is rotated are dependent on (1) the nature of the compound, (2) the concentration of the solution, (3) the path length through which the light travels, (4) the solvent, (5) the wavelength of light and (6) the temperature.

These variables are connected by the formula

$$\left[\alpha\right]_{\lambda}^{t} = \frac{\alpha}{l \times c} \quad \text{for water}$$

where $[\alpha]$ = specific rotation of the compound; t = temperature at which the experiment is carried out (usually constant at 20°C); and λ = wavelength of light used, usually the sodium D-line (589 nm). α = observed rotation. It is important to note the direction in which the light is rotated. If the light is rotated to the right, the compound is dextro-rotatory and the direction is denoted by

(+); and if the light is rotated to the left, the compound is laevorotatory and the direction is denoted by (−). l = path length through which the light travels through the solution under examination. It is measured in decimetres (dm) and in most polarimeters is 2 dm. c = concentration of the solution in g/ml.

The specific rotation of a compound under defined conditions of temperature and wavelength of light used is a standard physical constant and as such can be used in the identification or in the determination of concentration of that compound.

The Polarimeter

A diagrammatic representation of the essential parts of the polarimeter is given in Figure 2.8.

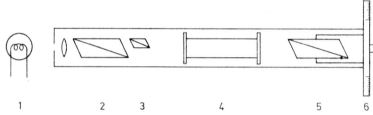

Figure 2.8. A polarimeter. 1, Light source of monochromatic light, usually a sodium discharge lamp. 2, A polariser, a Nicol prism or Polaroid, which produces plane-polarised light. 3, A device for rotating the transmitted light by about 10°; it is usually a small Nicol prism. This device splits the emergent light into two distinct zones; this enables the end-point of minimum light transmission to be more accurately obtained. 4, A polarimeter tube containing the solution under examination. 5, An analyser which is similar to the polariser. It can be rotated through 90° in either direction. 6, A scale connected to the analyser from which the degree of rotation is read

The observed emergent light is divided into two distinct semicircular zones. Examples of this are given in Figure 2.9.

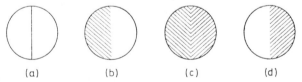

Figure 2.9. Emergent light through a polarimeter. (a) Maximum emergent light: the analyser is in line with the polariser. (b) Half-emergent light: the analyser is at 45° to the polariser. (c) Minimum emergent light: the analyser is crossed (at 90°) with the polariser. This position is taken as the end-point. (d) Half-emergent light: the analyser is at 45° (in the opposite direction from (b)) to the analyser

Mutarotation of Sugars

Glucose and other aldoses exist as internal hemiacetals; this formation produces another asymmetric carbon atom and therefore glucose exists in two forms: the α-form and the β-form. Glucose recrystallised from water exists in the α-form, which in solution slowly undergoes mutarotation to the equilibrium mixture (a mixture of the α- and β-forms with traces of the free aldehyde). The rate of mutarotation, which is catalysed by traces of alkali, can be followed by polarimetry.

Reagents and Apparatus

D-glucose (anhydrous)
1 M Na_2CO_3
Polarimeter
Sodium vapour lamp

Procedure

Switch on the sodium vapour lamp and allow 10 min for the lamp to warm up. Thoroughly rinse the polarimeter tube with distilled water and then fill the tube with water. Screw on the brass end-cap and ensure that the tube contains no air-bubbles. With the sodium lamp in position, rotate the analyser until the light intensity in the two semicircular zones is of equal darkness. Note the reading on the scale; this is the *zero* of the instrument. Repeat the zero determination.

Prepare a fresh solution of 10% w/v glucose in a 100-ml volumetric flask. Fill the polarimeter tube with this solution and determine the rotation by rotating the analyser until the two zones are of equal darkness. Note the reading and direction on the scale and subtract the zero reading from this reading. Determine the rotation every 10 min for the first 30 min and then every 20 min until a constant reading is obtained. Tabulate your results.

Prepare another fresh solution of 10% w/v glucose in a 100-ml volumetric flask. Just before the solution is up to the mark, add three drops of 1 M Na_2CO_3, shake well and make up to the mark. Fill the polarimeter tube with this solution and determine the rotation every 2 min for the first 10 min and then every 5 min for the next 20 min. Tabulate your results.

Plot two graphs of observed rotation against time and determine the specific rotation for α-D-glucose and the equilibrium mixture.

EXPERIMENT 8
PERIODATE OXIDATION

Periodate (IO_4^-) oxidises:
1. compounds with adjacent hydroxy groups
2. α-hydroxy carbonyl compounds
3. compounds with adjacent carbonyl groups
4. compounds with adjacent primary amine groups
5. α-amino alcohols
6. α-amino carbonyl compounds

During the oxidation, cleavage of the C—C bond between these groups occurs. Some examples are given below:

$$\begin{array}{c} |\\ -C-OH\\ |\\ -C-OH\\ | \end{array} + IO_4^- \longrightarrow \begin{array}{c} |\\ -C=O\\ +\\ -C=O\\ | \end{array} + IO_3^- + H_2O$$

$$\begin{array}{c} H\\ \diagdown C{=}O\\ |\\ -C-OH\\ | \end{array} + IO_4^- \longrightarrow \begin{array}{c} H.CO_2H\\ +\\ -C=O\\ | \end{array} + IO_3^-$$

$$\begin{array}{c} |\\ -C-OH\\ |\\ C=O\\ |\\ -C-OH\\ | \end{array} + IO_4^- \longrightarrow \begin{array}{c} |\\ -C=O\\ +\\ CO_2\\ +\\ -C=O\\ | \end{array} + IO_3^- + H_2O$$

Compounds with three adjacent hydroxy groups are oxidised by IO_4^- to yield a molecule of two different aldehydes and a molecule of formic acid:

$$\begin{array}{c} |\\ H-C-OH\\ |\\ H-C-OH\\ |\\ H-C-OH\\ | \end{array} + 2IO_4^- \longrightarrow \begin{array}{c} |\\ H-C=O\\ +\\ H\,CO_2H\\ +\\ H-C=O\\ | \end{array} + 2IO_3^- + H_2O$$

IO_4^- oxidation is used in the determination of structure of sugars. The mode of reaction distinguishes between aldo-sugars and keto-sugars.

CARBOHYDRATES

Consider the oxidation of an aldohexose in the pyranose form:

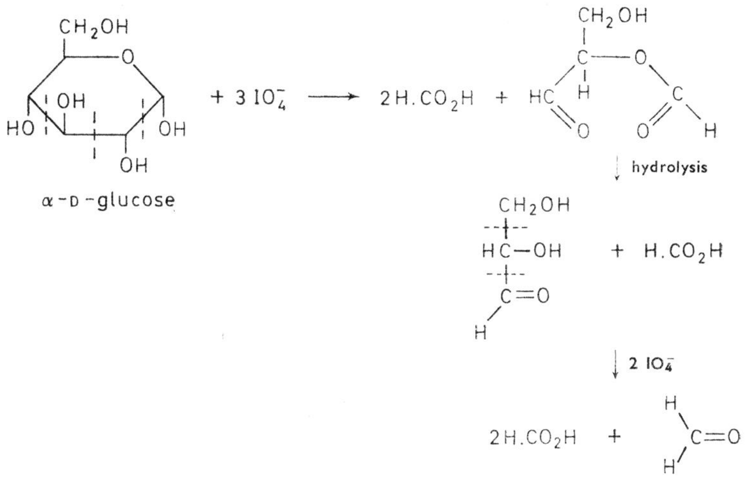

$$\alpha\text{-D-glucose} + 3\,IO_4^- \longrightarrow 2\,H.CO_2H + \text{(dialdehyde-formate intermediate)}$$

↓ hydrolysis

$$\begin{array}{c} CH_2OH \\ | \\ HC-OH \\ | \\ C=O \\ | \\ H \end{array} \quad + \quad H.CO_2H$$

↓ $2\,IO_4^-$

$$2\,H.CO_2H \quad + \quad \begin{array}{c} H \\ \diagdown \\ C=O \\ \diagup \\ H \end{array}$$

In this case 5 mol of IO_4^- is consumed (5 C—C bonds are broken) and 5 mol of H.COOH is formed. 3 mol of IO_4^- is consumed rapidly, which corresponds to the oxidation of the four adjacent hydroxy groups in the compound, while the last 2 mol of IO_4^- is consumed slowly, which corresponds to the oxidation between carbons 4, 5 and 6. The latter oxidation does not occur until the formate ester linked to C_5 has been hydrolysed.

Consider the oxidation of a ketohexose in the furanose ring structure:

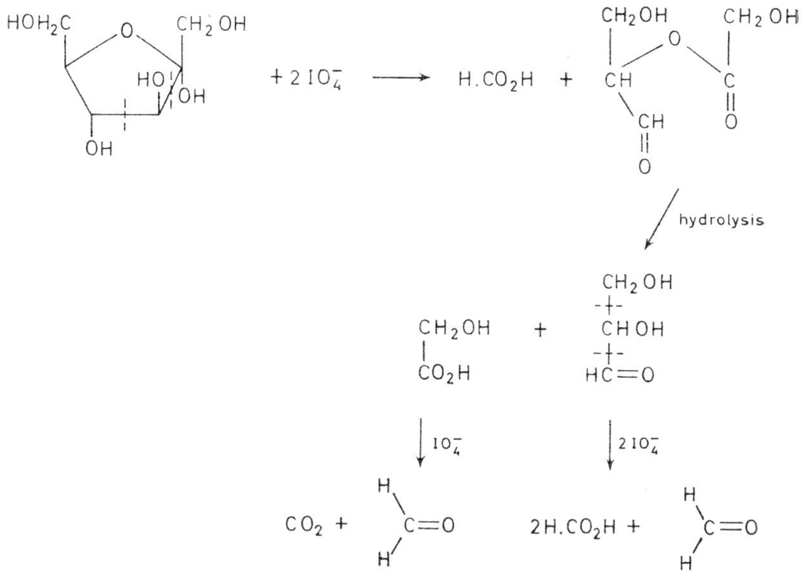

$$\text{(ketohexose furanose)} + 2\,IO_4^- \longrightarrow H.CO_2H + \text{(intermediate)}$$

↓ hydrolysis

$$\begin{array}{c} CH_2OH \\ | \\ CO_2H \end{array} \quad + \quad \begin{array}{c} CH_2OH \\ -|- \\ CHOH \\ -|- \\ HC=O \end{array}$$

↓ IO_4^- ↓ $2\,IO_4^-$

$$CO_2 + \begin{array}{c} H \\ \diagdown \\ C=O \\ \diagup \\ H \end{array} \qquad 2\,H.CO_2H + \begin{array}{c} H \\ \diagdown \\ C=O \\ \diagup \\ H \end{array}$$

In this case 5 mol of IO_4^- is consumed and 3 mol of $H.COOH$ is formed.

The ratio

$$\frac{\text{moles } H.COOH \text{ formed}}{\text{moles } IO_4^- \text{ consumed}}$$

is 1 in the oxidation of aldo-sugars and is < 1 in the oxidation of keto-sugars.

The amount of IO_4^- consumed is determined by the liberation of I_2 from KI in acid conditions. The I_2 is estimated by titration against standard thiosulphate.

During the reaction IO_4^- is reduced to IO_3^-:

1 mol IO_4^- liberates 4 mol I_2 according to the equation

$$IO_4^- + 7I^- + 8H^+ \rightarrow 4I_2 + 4H_2O$$

1 mol IO_3^- liberates 3 mol I_2 according to the equation

$$IO_3^- + 5I^- + 6H^+ \rightarrow 3I_2 + 3H_2O$$

The amount of I_2 liberated decreases as IO_4^- is reduced; a reduction of 1 mol of IO_4^- corresponds to a decrease of 1 mol of I_2 liberated.

The formic acid produced is determined by titration against a strong alkali, phenolphthalein being used as indicator. Ethylene glycol is added to the titration flask to react with excess periodate.

Reagents

D-fructose
D-ribose
10% w/v KI
0·5 M sodium periodate
Approx. 1 M H_2SO_4
Standardised sodium thiosulphate solution (approx. 0·1 M)
Standardised sodium hydroxide solution (approx. 0·1 M)
Phenolphthalein
1% w/v soluble starch
Ethylene glycol

Procedure

Weigh accurately 200–300 mg of D-fructose. Dissolve the sugar in about 5 ml of water and transfer the solution and washings to a 50-ml volumetric flask. Pipette 25 ml of 0·5 M $NaIO_4$ solution to

CARBOHYDRATES 45

the flask and make up to the mark. Prepare a blank containing 25 ml of water and 25 ml of 0·5 M $NaIO_4^-$ solution in a 50-ml volumetric flask. Leave the flasks in the dark for one week.

After one week, pipette 1-ml aliquots of reaction mixture and blank into two conical flasks. Add 10 ml of 10% KI and about 3 ml of 1 M H_2SO_4 to each flask. Titrate the liberated I_2 against standardised $Na_2S_2O_3$. A few drops of 1% soluble starch may be added near the end-point as an indicator. (Iodine should be titrated as soon as it is liberated, i.e. when the acid is added.) Carry out the titrations in triplicate.

Pipette, in triplicate, 5-ml aliquots of reaction mixture into a conical flask. Add 1 ml of ethylene glycol to react with the excess IO_4^- and allow the flask to stand for about 10 min, shaking occasionally. Titrate against standardised NaOH (preferably in a 10-ml burette), using phenolphthalein as indicator.

Repeat the experiment with D-ribose.

Calculation

MILLIMOLES OF IO_4^- CONSUMED

Let

molarity of $Na_2S_2O_3$	$= y$ M
titre value of blank (1-ml aliquot)	$= a$ ml
titre value of sample (1-ml aliquot)	$= b$ ml
weight of sugar	$= x$ mg in 50-ml sample

$$2 \text{ mol } Na_2S_2O_3 \equiv 1 \text{ mol } I_2$$

$$(a-b) \text{ ml} \times y \text{ M } Na_2S_2O_3 \equiv \frac{(a-b) \times y}{2} \text{ mmol } I_2$$

$\equiv$ decrease in I_2 liberated in 1-ml aliquot

Decrease in mmol I_2 liberated $\equiv$ mmol IO_4^- consumed

$$\frac{\text{mmol } IO_4^- \text{ consumed in}}{\text{1-ml aliquot}} = \frac{(a-b) \times y}{2}$$

$$\frac{\text{mmol } IO_4^- \text{ consumed in 50-ml}}{\text{sample } (x \text{ mg sugar})} = 50 \cdot \frac{(a-b) \times y}{2}$$

MILLIMOLES OF H.COOH FORMED

Let

    molarity of NaOH $= z$ M

    titre value (5-ml aliquot) $= d$ ml

$$1 \text{ mol NaOH} \equiv 1 \text{ mol H.COOH}$$

$$d \text{ ml (titre)} \times z \text{ M NaOH} \equiv d \times z \text{ mmol H.COOH}$$
$$= \text{amount of H.COOH formed in 5-ml aliquot}$$

$$\frac{\text{mmol H.COOH formed in}}{\text{50-ml sample } (x \text{ mg sugar})} = 10.(d \times z)$$

Calculate the ratio

$$\frac{\text{mmol H.COOH formed}}{\text{mmol IO}_4^- \text{ consumed}}$$

for (a) D-fructose, and (b) D-ribose. Comment on the results.

EXPERIMENT 9
GLYCOGEN LEVELS IN RAT LIVERS

Glycogen in this experiment is extracted from rat liver where (a) the rat has been fed on a normal diet containing carbohydrate, and (b) the rat has been starved for 48 h prior to the beginning of the experiment.

The tissue is disintegrated by refluxing at 100°C with 30% KOH; this treatment destroys proteins and reducing sugars but does not significantly destroy the glycogen. The glycogen is then precipitated with ethanol, the precipitate is hydrolysed with acid and the resulting glucose is estimated colorimetrically.

This experiment could conveniently be carried out by two pairs of students, one of whom disintegrates the liver of a normally fed rat, while the other disintegrates the liver of a starved rat. After disintegration, the pairs would exchange samples of the different liver digests.

Reagents and Apparatus

Rats which have been normally fed
Rats which have been starved for 48 h
30% w/v KOH
95% v/v ethanol
0·6 M HCl
2 M NaOH
Saline (0·9% NaCl)
0·1% w/v NaCl
Standard glucose solution containing 1 mg/ml
* 3, 5-Dinitrosalicylic acid reagent
Colorimeter

Procedure

ISOLATION OF GLYCOGEN

Decapitate the rat, remove the liver and drop it into 50 ml of cold saline in a 100-ml measuring cylinder. Roughly weigh the liver by displacement of saline. (The liver of a fed rat weighs about 8–10 g and that of a starved rat about 4–6 g.) Decant the saline, blot the liver dry on filter paper and then cut it into small pieces. Transfer the finely divided rat liver to a boiling tube and add 2 ml of 30% KOH, using a safety pipette, for each gramme of tissue. (The procedure up to adding the caustic potash should be carried out as quickly as possible to limit glycolysis.) Fit the tube with an air condenser and reflux the mixture in a boiling water-bath until the tissue is completely disintegrated (30–40 min).

Carefully transfer the mixture to a 50-ml centrifuge tube and centrifuge for 3 min. Decant the supernatant into a clean boiling tube. Using a safety pipette, pipette four 3-ml samples of liver digest from the fed rat and two 6-ml samples of digest from the starved rat into clean 50-ml centrifuge tubes. The samples from the fed rat are equivalent to 1 g of original tissue, and those from the starved rat to 2 g of original tissue. (The pairs of students working together would exchange two samples from the fed rat and one sample from the starved rat, so that each pair has three samples—two from the fed rat and one from the starved rat. Each pair would then continue independently.)

For each millilitre of digest add 1·2 vols. of 95% ethanol. Mix well with a fine glass rod, fit air condensers to the tubes and place in a boiling water-bath. Remove the tubes as soon as the ethanol begins to boil. Cool thoroughly and then centrifuge for 3 min. Decant off the supernatant from the precipitated glycogen, finally inverting the tube to remove as much excess liquid as possible. Add 2 ml of 0·1% NaCl to each tube and dissolve the glycogen by stirring with a glass rod. Add 1 ml of 95% ethanol for each millilitre of original digest, then mix and centrifuge for 3 min. Decant off the supernatant as before. Expel the residual ethanol by heating in a boiling water-bath for about 2 min.

HYDROLYSIS AND ESTIMATION OF GLYCOGEN

Add 5 ml of 0·6 M HCl to each sample. Fit tubes with air condensers and reflux mixtures in a boiling water-bath for 2–3 h.

Quantitatively transfer hydrolysate of liver glycogen from the fed rat into a 50-ml volumetric flask and hydrolysate of liver glycogen from the starved rat into a 25-ml volumetric flask. Add water until the flasks are about 3/4 full. Adjust the pH to 6 by dropwise addition of 2 M NaOH using universal indicator paper. Make the solutions up to the mark.

Prepare a glucose standard curve by pipetting, in duplicate, the following reagents into a series of labelled test-tubes:

Tube	Standard glucose solution (ml)	Water (ml)	3 : 5 dinitrosalicylic acid reagent (ml)
1	0	1·0	2·0
2	0·2	0·8	2·0
3	0·4	0·6	2·0
4	0·7	0·3	2·0
5	1·0	0	2·0

Place the tubes in a boiling water-bath and boil for exactly 5 min. Thoroughly cool and add 7 ml of water to each tube. Read absorbance against reagent blank at 540 nm (green filter). Plot a standard glucose curve.

Repeat the above experiment but add (a) diluted glycogen hydrolysate from the fed rat and (b) diluted glycogen hydrolysate from the starved rat, instead of standard glucose solution. Carry out glucose determinations of glycogen hydrolysates in duplicate.

Calculation

From the standard glucose curve determine the glucose in mg/ml present in the diluted glycogen hydrolysate. Hence, determine the glucose in mg/g originally present in the liver. Since 162 parts of glycogen and 18 parts of water yield 180 parts of glucose, the glucose found should be multiplied by $162/180 = 0.90$ to give glycogen.
About 3% glycogen is destroyed during hydrolysis. Therefore, the glycogen content is more accurately expressed by multiplying the glucose by 0·93.
Express the results as g glycogen/100 g liver.

EXPERIMENT 10
SEPARATION OF SUGARS BY PAPER CHROMATOGRAPHY

Chromatography is a technique for separating closely related groups of compounds. The separation is brought about by differential migration along a porous medium and the migration is caused by the flow of solvent. Within limits chromatography can be divided into two types: partition chromatography and adsorption chromatography.

Paper chromatography is an example of liquid/liquid partition chromatography. In this type of chromatography separation is due to the differential partition of solutes between two liquid phases. One liquid phase is bound to the porous medium—for example, the 8–12% water bound in the cellulose of paper; this phase is referred to as the stationary phase. The other liquid phase, the mobile phase, flows along the porous medium. As the mobile phase flows over the solute mixture, the individual solutes partition themselves between the aqueous stationary phase and the organic mobile phase relative to their solubilities in the two phases. The more soluble an individual solute is in the mobile phase, the faster it will travel along the paper; and, conversely, the more soluble a solute is in the aqueous stationary phase, the slower it will travel.

Paper chromatography can only be used for groups of compounds which are soluble or partially soluble in water. Furthermore, the mobile phase must be a mixture in which the compounds to be separated are soluble or partially soluble.

In general, the migration of solutes takes place while the porous medium is in an atmosphere saturated with the solvent. Chromatography tanks are used for this purpose.

In paper chromatography the solute or solute mixture is spotted in solution along a base line on a sheet of filter paper. The mobile phase is allowed to flow over the spots either ascending the paper by capillary action or descending the paper by gravity. The former technique is referred to as ascending chromatography and the latter is referred to as descending chromatography (see Figure 2.10).

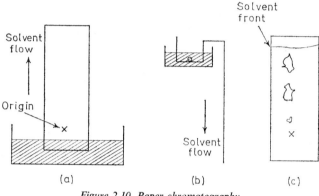

Figure 2.10. Paper chromatography

Compounds are characterised by their relative rates of flow with respect to the solvent front, which is referred to as the R_f value:

$$R_f = \frac{\text{distance from origin travelled by substance}}{\text{distance from origin travelled by solvent}}$$

The R_f value of a substance is constant under identical conditions of the experiment, e.g. the nature of the solvent mixture, temperature, pH, pressure, ascending or descending chromatography, etc. Therefore an unknown can often be identified by its R_f value or by its R_f values in two solvent systems. On account of the large number of variables, it is the practice for one or more known substances called markers to be run with the unknown mixture. The distance travelled by the unknowns compared with the markers provides a means of identification of the unknowns.

Separation of a mixture of dyes or pigments can readily be followed during the development of the chromatogram. If, on the other hand, the constituents of the mixture are colourless, then they are detected chemically by a suitable locating agent, after the chromatogram has been developed.

Paper chromatography can also be used for quantitative work. A known amount of the mixture is spotted on the paper with a calibrated micro-syringe; and after the chromatogram has been developed, the individual spots are eluted from the paper and subsequently estimated by a suitable method.

Reagents and Apparatus

1% w/v solutions of glucose, maltose, galactose, lactose, fructose, sucrose, soluble starch and brown sugar
Solution of hydrolysed starch
Solution of hydrolysed brown sugar
* Solution of fat-free deproteinised milk and solution of hydrolysed fat-free deproteinised milk
10% w/v solution of sodium tungstate, 1/3 M H_2SO_4
Solvent (ethyl acetate: pyridine : water 110 : 50 : 40 by volume)
* Locating agent: Aniline–diphenylamine reagent
Cylindrical chromatography tank
Dipping tray
Hair-dryer

Procedure

Prepare 50 ml of solvent mixture and place in the tank. Seal the lid and allow the solvent to saturate the tank for 30 min.
Prepare the hydrolysed carbohydrate solutions and the protein-free milk solution as follows:

1. Add 1 ml of 2 M HCl to 5 ml of 1% starch solution and to 5 ml of 1% brown sugar solution. Heat the mixtures in a boiling water-bath for 15 min. Cool the solutions and neutralise them to about pH 7 using pH paper with 2 M NaOH. Label the solutions 7 and 8, respectively.
2. To 2 ml of fat-free milk add 4 ml of 10% sodium tungstate; then add, slowly and with constant shaking, 4 ml of 1/3 M H_2SO_4. Allow the mixture to stand for 5 min. Filter the mixture through a Whatman No. 42 filter paper. To 2 ml of the filtrate add 0·5 ml of 2 M HCl and heat in a boiling water-bath for 15 min. Cool and neutralise to about pH 7 using pH paper with 2 M NaOH. Label the filtrate and the hydrolysed filtrate 9 and 10, respectively.

Prepare a sheet of Whatman No. 1 filter paper, 30×30 cm, by drawing a base line 3 cm from the bottom, and mark 10 origins (numbered 1–10) $2\frac{1}{2}$ cm apart. (Do not touch the rest of the paper because of grease on your hands.)

On origins 1–6 spot, using a capillary pipette or platinum loop, three drops of 1% solutions of glucose, maltose, galactose, lactose, fructose and sucrose, respectively. On origins 7, 8 and 10 spot three drops of hydrolysed starch solution, hydrolysed sugar solution and hydrolysed protein-free milk. On origin 9 spot three drops of protein-free milk. (After addition of the first drop, allow the spots to dry before adding the second and third drops, so that the area of the spots is kept as small as possible.)

Staple the paper into a cylinder and place the cylinder with the spotted end down in the tank, taking care not to let the paper touch the glass walls. Allow the chromatogram to run until the solvent front approaches the top of the paper (at least 3 h). Remove the chromatogram, mark the solvent front with a pencil and then dry the chromatogram with a hair dryer. Pour the aniline–diphenylamine locating agent into a dipping-tray and dip the chromatogram. Heat the dipped chromatogram for 5 min at 80°C to develop the spots.†

Outline the carbohydrate spots. Record their R_f values. Comment on the constituents of starch and sugar and the carbohydrate of milk.

---

† With this locating agent, sucrose and fructose give a brown spot and the other sugars used in this experiment give blue spots.

CHAPTER 3

Amino Acids and Proteins

STRUCTURE OF AMINO ACIDS

Amino acids are the building blocks of proteins. There are about 22 naturally occurring α-amino acids which occur frequently in proteins (see Table 3.1). Their general structure is:

$$\underset{\underset{NH_2}{|}}{R-CH-CO_2H} \rightleftharpoons \underset{\underset{NH_3^+}{|}}{R-CH-CO_2^-}$$
Zwitterion form

Amino acids are ampholytes, since they possess both an acidic and a basic group. They exist in solution as charged molecules. At their isoelectric point, the pH at which the molecule possesses no net charge, they exist in the zwitterion form or dipolar ion form; at a pH < isoelectric point they exist as cations, and at a pH > isoelectric point they exist as anions.

At their isoelectric point amino acids are least soluble in water and do not migrate to either electrode under the influence of an electric current.

Except when R is a hydrogen atom, all the groups attached to the α-carbon atom will be different and therefore the α-carbon atom will be asymmetric. In such cases, the amino acids exist in two optically active forms. Most naturally occurring amino acids are in the L-series, derived from L-glyceraldehyde, the reference compound, e.g.

L-glyceraldehyde

$$\begin{array}{c} H\diagdown \;\;\diagup O \\ C \\ | \\ HO-C-H \\ | \\ CH_2OH \end{array}$$

L-alanine

$$\begin{array}{c} HO\diagdown \;\;\diagup O \\ C \\ | \\ NH_2-C-H \\ | \\ CH_3 \end{array}$$

Table 3.1

Amino acid	Shortened name	Isoelectric point	Formula
1. Aliphatic monoamino monocarboxylic amino acids			
Glycine	Gly	6·0	CH_2CO_2H — NH_2
Alanine	Ala	6·0	$CH_3.CH.CO_2H$ — NH_2
Valine	Val	6·0	$CH_3.CH.CH.CO_2H$ / $H_3C\ NH_2$
Leucine	Leu	6·0	$CH_3.CH.CH_2.CH.CO_2H$ / $CH_3\ \ NH_2$
Isoleucine	Ileu	6·0	$CH_3.CH_2.CH.CH.CO_2H$ / $H_3C\ NH_2$
2. Hydroxy-containing amino acids			
Serine	Ser	5·7	$CH_2.CH.CO_2H$ / $OH\ NH_2$
Threonine	Thr	5·6	$CH_3.CH.CH.CO_2H$ / $OH\ NH_2$
3. Sulphur-containing amino acids			
Cysteine	Cys	5·1	$HS-CH_2.CH.CO_2H$ — NH_2
Cystine	Cys	4·8	$S-CH_2-CH-CO_2H$ / $\ \ \ \ \ \ NH_2$ / $S-CH_2-CH.CO_2H$ / $\ \ \ \ \ \ NH_2$
Methionine	Met	5·7	$CH_3.S.CH_2.CH.CO_2H$ — NH_2
4. Monoamino dicarboxylic amino acids (and amides)			
Aspartic acid	Asp	2·8	$CO_2H.CH_2.CH.CO_2H$ — NH_2

AMINO ACIDS AND PROTEINS

Table 3.1 *(cont.)*

Amino acid	Shortened name	Isoelectric point	Formula
Asparagine	Asn		$NH_2-\underset{\underset{O}{\|}}{C}.CH_2.\underset{\underset{NH_2}{\|}}{CH}.CO_2H$
Glutamic acid	Glu	3·2	$CO_2H.CH_2.CH_2.\underset{\underset{NH_2}{\|}}{CH}.CO_2H$
Glutamine	Gln		$NH_2-\underset{\underset{O}{\|}}{C}.CH_2.CH_2.\underset{\underset{NH_2}{\|}}{CH}.CO_2H$

5. Diamino monocarboxylic amino acids

Lysine	Lys	10·0	$NH_2.(CH_2)_4.\underset{\underset{NH_2}{\|}}{CH}.CO_2H$
Arginine	Arg	10·8	$\underset{NH_2}{\overset{NH_2}{\diagdown}}C=N-(CH_2)_3.\underset{\underset{NH_2}{\|}}{CH}.CO_2H$

6. Cyclic and aromatic amino acids

Phenylalanine	Phe	5·5	$C_6H_5-CH_2.\underset{\underset{NH_2}{\|}}{CH}.CO_2H$
Tyrosine	Tyr	5·7	$HO-C_6H_4-CH_2.\underset{\underset{NH_2}{\|}}{CH}.CO_2H$
Tryptophan	Try	5·9	indole—$CH_2.\underset{\underset{NH_2}{\|}}{CH}.CO_2H$
Histidine	His	7·6	imidazole—$CH_2.\underset{\underset{NH_2}{\|}}{CH}.CO_2H$
Proline	Pro	6·3	pyrrolidine—CO_2H
Hydroxyproline	Hyp	5·8	hydroxypyrrolidine—CO_2H

PROPERTIES AND REACTIONS OF AMINO ACIDS

1. Amino Acids as Buffers at Certain pH Values

$$\underset{\text{Base}}{\begin{array}{c} NH_2 \\ | \\ CH_2 \\ | \\ CO_2^- \end{array}} \underset{H^+}{\rightleftharpoons} \underset{\text{Salt}}{\begin{array}{c} \overset{+}{NH_3} \\ | \\ CH_2 \\ | \\ CO_2^- \end{array}} \underset{H^+}{\rightleftharpoons} \underset{\text{Acid}}{\begin{array}{c} \overset{+}{NH_3} \\ | \\ CH_2 \\ | \\ CO_2H \end{array}}$$

In acid conditions an amino acid exists as a cation, which acts as an acid, and in alkaline conditions it exists as an anion, which acts as a base. At its isoelectric point an amino acid is in the zwitterion or salt form; this form possesses no buffering action at all and addition of small amounts of H^+ or OH^- ions rapidly alters the pH of the solution. When the amino acid exists as a mixture of the salt and acid or base form in approximately equimolar concentrations (in the region of its pK values), the system acts as a stable buffer.

2. Reaction with Nitrous Acid

Nitrous acid reacts with primary aliphatic amino groups with the liberation of nitrogen to form alcohols:

$$\begin{array}{c} R-CH-CO_2H \\ | \\ NH_2 \end{array} + HNO_2 \longrightarrow \begin{array}{c} R-CH-CO_2H \\ | \\ OH \end{array} + N_2\uparrow + H_2O$$

This reaction can be used to estimate amino acids quantitatively by measuring the volume of nitrogen liberated, as 1 mol liberates 22·4 litres of N_2 at STP. This is the basis of Van Slyke's method of estimating amino-nitrogen.

3. Reaction with Ninhydrin

All amino acids react with ninhydrin to give a characteristic purple colour (except proline and hydroxyproline, which give a yellow colour). Ninhydrin is reduced and the amino acid is deaminated

AMINO ACIDS AND PROTEINS

and decarboxylated. One molecule of reduced ninhydrin reacts with one molecule of ammonia and one molecule of unreacted ninhydrin to form a purple complex:

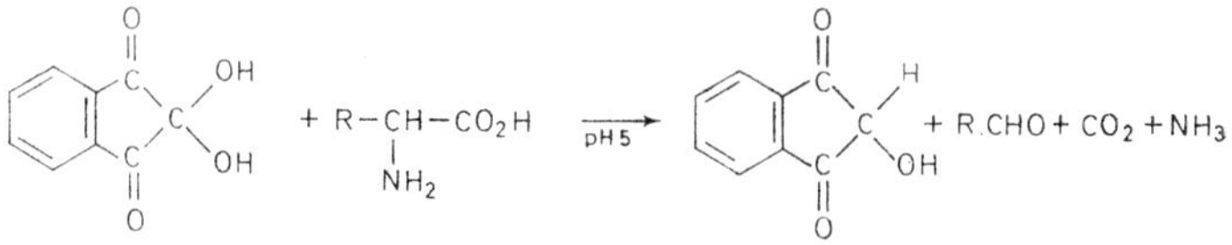

[Reaction scheme: Triketohydrindene hydrate (ninhydrin) + R—CH(NH$_2$)—CO$_2$H →(pH 5) Reduced ninhydrin + R.CHO + CO$_2$ + NH$_3$]

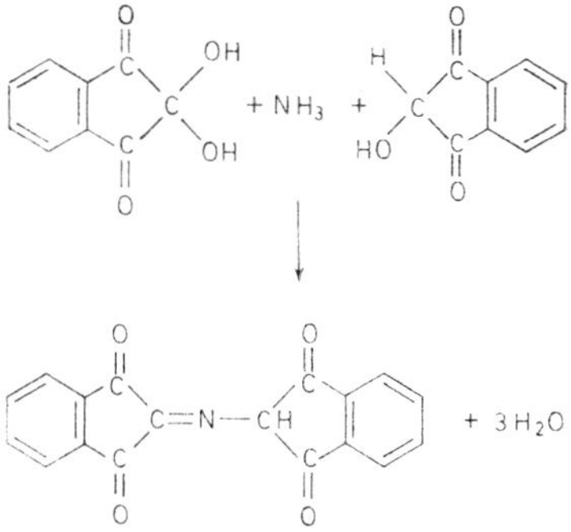

[Reaction scheme: Reduced ninhydrin + NH$_3$ + ninhydrin → Purple complex + 3 H$_2$O]

4a. Reaction with 1-fluoro-2,4-dinitrobenzene (DNB)

The amino group reacts with DNB in alkaline conditions to form a yellow derivative:

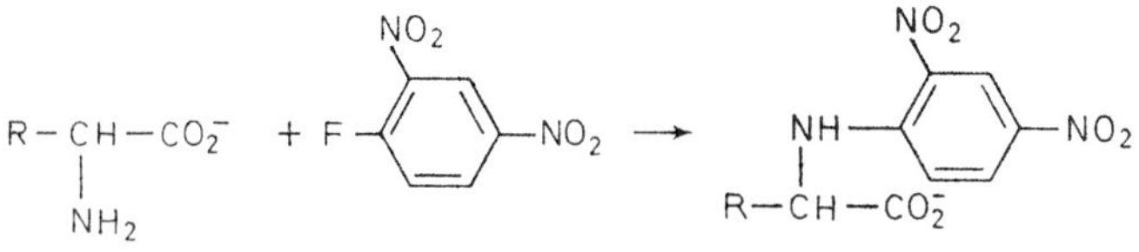

[Reaction: R—CH(NH$_2$)—CO$_2^-$ + F—C$_6$H$_3$(NO$_2$)$_2$ → NH—C$_6$H$_3$(NO$_2$)$_2$ attached to R—CH—CO$_2^-$ — Yellow DNB derivative]

This reaction is important in determining the terminal amino group in proteins.

4b. Reaction with Dansyl Chloride (1-dimethylaminonaphthalene-5-sulphonyl chloride) (DNS)

A free amino group reacts with dansyl chloride to form a fluorescent compound:

$$\underset{\text{Fluorescent DNS derivative}}{\text{[1-(dimethylamino)naphthalene-5-sulfonyl chloride]}} + \text{R.CH(NH}_2\text{)CO}_2\text{H} \longrightarrow \text{[1-(dimethylamino)naphthalene-5-SO}_2\text{-NH-CH(R)-CO}_2\text{H]} + \text{HCl}$$

This reaction is now extensively used for amino end-group analysis on a micro scale.

5. Reaction with Mineral Acids

The basic group reacts with mineral acids to form crystalline salts:

$$\underset{\text{NH}_2}{\text{R.CH-CO}_2\text{H}} + \text{HCl} \longrightarrow \underset{\underset{\text{Amino acid hydrochloride}}{\text{NH}_3^+ \text{ Cl}^-}}{\text{R-CH-CO}_2\text{H}}$$

6. Formol Titration

The carboxylic group is neutralised by alkalis but the amino group interferes with a straight titration. If formaldehyde is added, it reacts with the amino group (aldehyde addition-type reaction), allowing the amino acid to be titrated directly against an alkali.

7. Colour Reactions of Amino Acids

Many amino acids, either free or bound in a protein, give characteristic colour reactions (see Experiment 11).

PROTEIN STRUCTURE

Proteins are made up from about 22 amino acids and vary in molecular weight from 5000 to 20 000 000. Compared with polysaccharides, which are made up from one or two building blocks, it can be seen that proteins are much more complex. The number of permutations and combinations which can be obtained from 22 amino acids is enormous and therefore the possible number of proteins is very large indeed. Moreover, the function of proteins is more varied than any other type of naturally occurring macromolecule, in that they act as enzymes (biological catalysts), hormones,

AMINO ACIDS AND PROTEINS 59

antibodies, carriers of O_2 and CO_2, storage molecules, structural units, energy sources, etc.

It is not only the sequence of amino acids which is important in protein structure; equally important for physiological activity is the three-dimensional structure. Protein structure is divided into four types, viz. primary, secondary, tertiary and quaternary.

Primary Structure

Primary structure is the amino acid sequence in proteins. Amino acids are condensed through the covalent peptide linkage

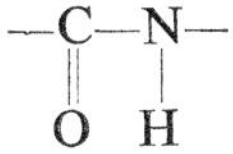

$$\begin{array}{c} -C-N- \\ \parallel \mid \\ O H \end{array}$$

to form polypeptides (see Figure 3.1).

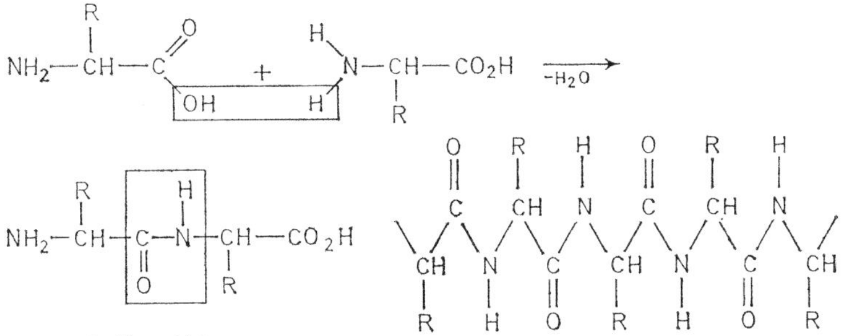

Figure 3.1. Primary structure of proteins: a polypeptide

Secondary Structure

Secondary structure is defined as all the types of conformation which polypeptide chains can take up. A protein consists of the sum of these parts.

HELIX FORMATION

The α-helix

The most important helical formation is the α-helix. The model deduced from X-ray data (Pauling and Corey, 1952) showed 18 amino acids per 5 turns; that is, 3·6 amino acids per turn.

This helix is stabilised by: (1) hydrogen bonding between the carbonyl and imino group, as shown in Figure 3.2; (2) salt bridge

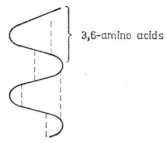

Figure 3.2. The α-helix stabilised by H-bonding

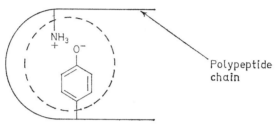

Figure 3.3. Electrostatic interactions: salt bridge formation

formation between the ionic side groups of basic, acidic and phenolic amino acids (see Figure 3.3); and (3) van der Waals' forces between similar hydrophobic side groups of amino acids (see Figure 3.4).

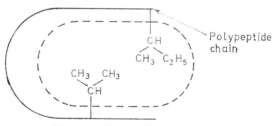

Figure 3.4. Van der Waals interactions

The β-helix

The β-helix is a parallel sheet type structure, as shown in Figure 3.5.

The forces stabilising this structure can be intermolecular or intramolecular. The parallel polypeptide chains may exist as α-helixes. Cross-linking between the parallel chains is mainly due to hydrogen bonding and disulphide bridges. This structure gives a protein a high degree of elasticity. Many fibrous proteins (for example, keratin) possess the β-helix structure.

Figure 3.5. The β-helix: parallel polypeptide chains

The triple-chain helix

Collagen exists as a triple-chain interwoven helix, each chain itself being a helix, as shown in Figure 3.6. This helix is stabilised by hydrogen bonding between the different chains.

Figure 3.6. A triple-chain helix

Collagen contains a high proportion of glycine and hydroxyproline.

The random coil

This type of structure is actually defined as the polypeptide chain being extended to the furthest possible extent. This type of structure, in practice, very rarely occurs. It is the amino acid composition of a polypeptide which determines the degree to which a chain can be extended, and a fully extended chain can only exist when the amino acid side groups are small, as in this case there is little stearic hindrance. In practice the random coil can be defined as the lack of any ordered system.

HYDROPHOBIC INTERACTION (LINDERSTRØM–LANG)

This has been compared to the emulsifying action of soap. A soap which has its hydrophobic groups orientated towards the centre of a globule and its hydrophilic groups along the surface stabilises an oil-in-water emulsion.

This arrangement of groups is similar in proteins and is referred to as the hydrophobic interaction. Thus it can be seen that the hydrophobic interaction is not a bond but a charge effect.

The hydrophobic interaction accounts for the compact shape of globular proteins. Water-soluble proteins have their hydrophobic groups orientated towards the centre of the molecule and their hydrophilic groups around the surface of the molecule. The arrangement of groups is the other way round in water-insoluble proteins.

DISULPHIDE BRIDGES

Cystine contains two amino acid residues linked together through a disulphide bond:

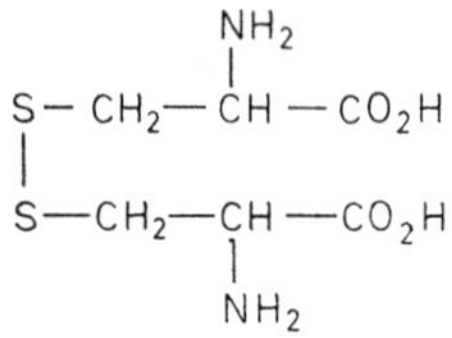

$$\begin{array}{c} \text{S}-\text{CH}_2-\overset{\overset{\displaystyle NH_2}{|}}{\text{CH}}-\text{CO}_2\text{H} \\ | \\ \text{S}-\text{CH}_2-\underset{\underset{\displaystyle NH_2}{|}}{\text{CH}}-\text{CO}_2\text{H} \end{array}$$

This amino acid often links two chains together or links a single chain in a definite configuration; this is represented in Figure 3.7.

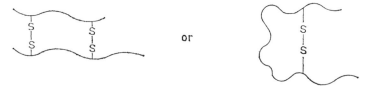

Figure 3.7. *The disulphide bridge formation*

THE AMINO ACIDS PROLINE AND HYDROXYPROLINE

These two amino acids are secondary amines, and when they form peptide linkages they possess no imino hydrogen to stabilise the helix by hydrogen bonding. Therefore whenever either of these

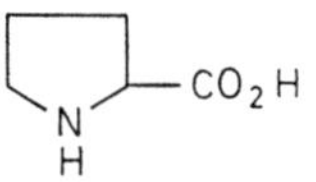

[Proline structure] —CO₂H Proline

amino acid residues occurs in the chain it causes a break in the helix (see Figure 3.8).

Figure 3.8. *A proline residue causing a kink in the helix*

Tertiary Structure

Tertiary structure is the interaction of all or some of the secondary structures to form a single three-dimensional unit (see Figure 3.9).

Figure 3.9. A hypothetical structure of a protein

Quaternary Structure

Some proteins are physiologically inactive as the single unit and the units combine to form dimers or trimers, etc., to become physiologically active proteins. The units are usually linked together by hydrogen bonding.

DENATURATION

Native proteins are highly ordered, complex molecules. Any force which alters the three-dimensional organisation within the molecule without rupturing the peptide bond causes denaturation. Denatured proteins have different properties from those of the native proteins and they no longer possess physiological activity. In addition, denaturation decreases the solubility of water-soluble proteins because the arrangement of the hydrophilic groups orientated towards the surface is destroyed. Common denaturing agents are heat, extremes in pH, oxidation and reduction which affects the disulphide bond, agitation, radiation and urea. Different proteins have different susceptibilities towards different agents. Denaturation can be a reversible process. Sometimes, when the agent is removed, as in the case of urea, the native protein is restored.

PRECIPITATION OF PROTEINS

Proteins are lyophilic (in particular, hydrophilic) colloids; they are stabilised by both charge and protein/solvent interaction. When one of these stabilising influences is removed, the protein is sometimes precipitated; when both of these influences are removed, the protein is always precipitated.

Isoelectric Point Precipitation

Proteins, like amino acids, are least soluble at their isoelectric point. Some proteins—for example, casein—are precipitated by adjusting the pH of the solution to the isoelectric point of the protein.

Salting-out

Addition of a neutral salt causes precipitation of proteins; different proteins are precipitated at different salt concentrations. Precipitation most readily occurs at the isoelectric point of the protein. The effect of the salt is to remove water, which decreases the protein/water interaction and thereby increases the protein/protein interaction, thus causing precipitation. Ammonium sulphate is most commonly used for salting-out proteins because it is extremely soluble and changes of temperature have little effect on its solubility.

Solvent Extraction

Addition of certain organic solvents, such as ethanol and acetone, precipitate proteins; and, again, precipitation occurs most readily at the isoelectric point of the protein. Organic solvents also remove water, thus decreasing the protein/water interaction. This procedure should be carried out at 0°C, otherwise denaturation may occur.

Complex Anions

Acids such as trichloracetic, sulphosalicylic, picric, tannic, tungstic, phosphomolybdic, etc., precipitate proteins. They are often referred to as protein precipitants. Precipitation is probably due to the neutralisation of the positive charge on the protein by the large anion (most proteins in neutral solution exist as anions but addition of acid converts them to cations). This type of precipitation often causes denaturation owing to an extreme in pH.

Heavy Metal Cations

Heavy metal cations such as Hg^{2+}, Pb^{2+}, Cu^{2+}, Ag^+, Au^+, Pt^{4+}, etc., precipitate proteins in alkaline conditions. The large cation destroys the negative charge on the protein, thus causing precipitation. These ions often denature the protein because they react with SH groups to form sulphides.

SEPARATION AND PURIFICATION OF PROTEINS

The separation and purification of a protein is essentially isolating the protein one requires from other substances and other proteins.

Purification is based on the physical and chemical properties of the protein, e.g. solubility, heat stability, charge, molecular size and shape, etc. No single extraction technique is sufficient to separate the protein one requires; separation is based on a series of fractionation procedures. Each stage isolates a group of proteins and each successive procedure reduces the number of proteins in the group.

The first priority is to find a specific assay for the protein one requires; the assay should be quick rather than precise in the preliminary stages of purification. After each fractionation procedure, one carries out the assay to ascertain (a) in which fraction the protein is—for example, the precipitate or the supernatant; (b) that the protein one requires has been concentrated; and (c) whether denaturation has occurred—this is indicated by loss of activity. If denaturation has occurred, one should try mixing together again the fractions separated in the last procedure; if activity returns, either the procedure has separated the protein into sub-units which are inactive separately or an essential co-factor has been separated from the protein. If mixing does not restore the activity, it means that the protein has been denatured and that particular technique cannot be used. It is important to remember that one purifies a specific activity, not a protein, and therefore if there are two or more proteins which behave similarly in the assay, one initially obtains a fraction in which these proteins occur together.

Most proteins are present in tissues to different extents; therefore one chooses a source which is rich in the protein one requires. However, it is still likely to be present in fairly low concentrations; therefore the initial procedures will be applied to a large amount of tissue. This means that the initial procedures will be applied to large volumes, and at each successive stage the volume of protein solution will be reduced. This factor usually governs the order in which the purification procedures are carried out. If a tissue is chosen as the source of protein, it is first homogenised, which produces a cell-free extract, and then centrifuged. The supernatant may be subjected to ultracentrifugation, which further separates it into subcellular fractions, as shown in Figure 3.10.

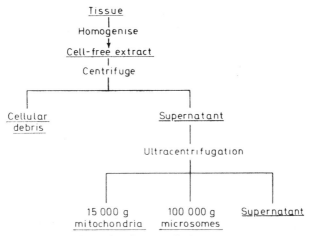

Figure 3.10. Flow diagram of subcellular fractions

An assay is carried out on each fraction to localise the protein. If the protein is present in a precipitate, the fraction is redissolved in a suitable amount of water or buffer. Also, the source of protein may be an extracellular fluid.

Stage 1

The initial step in separation usually depends on the proteins' solubilities in various concentrations of salt solutions.

Many proteins are precipitated with neutral salts, such as ammonium sulphate, or with organic solvents, such as ethanol or acetone. Different proteins are precipitated at different concentrations; the precipitate is successively removed at increasing salt or solvent concentration. The protein one requires should appear in one of these fractions or in the saturated solution. If the protein is present in a precipitate, the fraction is redissolved in a suitable amount of water or buffer.

Salt fractionation or solvent extraction is suitable as an initial separation technique, since it can be carried out on large volumes. These initial stages are usually carried out at low temperatures (about 4°C).

Stage 2

The next stage is often an adsorption technique. The proteins in solution are brought into contact with a highly absorbent material and selective adsorption takes place. The most frequently used adsorbent is calcium phosphate in the gel form. Adsorption on this medium occurs at low salt concentration and at a pH of around 5. Therefore, before this purification technique can be carried out, the protein solution must be dialysed to remove the small molecules.

Calcium phosphate gel is used by adding it to the protein solution and then stirring, centrifuging and decanting. If the protein one requires is adsorbed, it remains on the gel; if it is not adsorbed, it remains in the supernatant. Proteins are eluted from the gel at high salt concentrations and at approximately pH 7·6.

Calcium phosphate gel can also be used in a column together with an inert support, such as Celite, which makes the column permeable to the eluting reagent.

Stage 3

The next separation technique is often separation of proteins by charge differences at different pH values. This separation is achieved by ion exchange chromatography. The technique is based on the differential absorption of partially charged molecules, or ions in solution, to an ion exchange resin. The resins developed for this purpose must be porous and insoluble and carry an electrovalent charge. The charged molecules are eluted from the resin with an ionic solvent. Low concentrations of ions displace weakly bound molecules, and larger volumes or greater concentrations are needed to displace strongly bound molecules. The resin possesses a three-dimensional open lattice structure and the majority of exchangeable ions are held within this lattice structure.

In general, the larger the charge on the ion or molecule, and to a lesser extent the larger the size of the ion or molecule, the more firmly it is bound to the resin. The increase of affinity with ionic or molecular size occurs only when entry into the lattice structure is still possible. Large molecules, although highly charged, may not become bound and will be present in the effluent.

There are two types of ion exchange resins. These are cation exchange resins, which bind positively charged molecules, and anion exchange resins, which bind negatively charged molecules. Ion exchange resins are polymers which contain a suitable functional group carrying an electrovalent charge, such as a carboxyl, phenolic or sulphonic acid group in cation exchangers, and a tertiary or quaternary amine group in anion exchangers.

Synthetic exchange resins are made in the form of micro spheres or beads. They are either high molecular weight, cross-linked organic polymers, or cellulose, both types having a suitable functional group substituted in the polymer. The organic polymers are usually polystyrene cross-linked with divinyl benzene. The percentage of cross-linking agent added determines the critical molecular size for entry of a molecule into the lattice structure.

Mode of action:

Cation exchange

$$2\ RSO_3^-H^+ + Ca^{2+} \rightleftharpoons (RSO_3^-)_2Ca^{2+} + 2\ H^+$$
$\qquad\qquad$ Influent $\qquad\qquad\qquad\qquad$ Effluent

Anion exchange

$$R\overset{+}{N}(CH_3)_3OH^- + Cl^- \rightleftharpoons R\overset{+}{N}(CH_3)_3Cl^- + OH^-$$
$\qquad\qquad$ Influent $\qquad\qquad\qquad\qquad$ Effluent

Ion exchange celluloses are commonly used for protein separation. Those most used are DEAE (diethyl aminoethyl: —O—C_2H_4.N(C_2H_5)$_2$) cellulose and CM (carboxy methyl: —O—CH_2.COOH) cellulose. The former is an anion exchanger and is used to separate neutral and acidic proteins; the latter is a cation exchanger and is used to separate neutral and basic proteins. The degree of charge on the protein molecules determines the strength of binding in the column.

Stage 4

A further technique for separating proteins is based on their molecular sizes. This separation is achieved by gel filtration, which works in the opposite way to a kitchen sieve—it holds back the small molecules and lets the large molecules through. Large molecules cannot penetrate the gel particles and, hence, remain in the void volume outside the particles, and are eluted through the column at the same rate as the solvent. Small molecules, on the other hand, penetrate the water held inside the gel particles and they continually move into and out of the gel particles. Their degree of retardation in the column depends on the volume of water inside the particles relative to the volume of water outside the particles; the greater the volume inside the particles with respect to the volume outside, the slower the small molecules are eluted through the column. Molecules of intermediate sizes may penetrate only part of the gel water, and so may be separated from both the larger and smaller molecules. Such a separation requires a long column.

The most frequently used gel filtration media are Sephadexes, which are cross-linked dextrans. The greater the degree of cross-linking, the smaller the molecules it excludes. Polyacrilamide gel and starch gel also act as gel filtration media; these two media are mainly used in electrophoresis.

Electrophoresis

The principle of electrophoresis is described in Experiment 14. Proteins can be separated by zone electrophoresis using any of the porous media mentioned in that experiment. Electrophoresis readily ascertains the number of proteins present in a mixture. It separates small quantities of protein dissolved in a small volume of fluid; for this reason it is more useful as a criterion of purity than as a preparative method. Also, electrophoresis is often run in series with ion exchange chromatography or gel filtration; this shows the number of proteins present and therefore the number of elution peaks one should expect from the column.

The purified or partially purified protein obtained by the final separation can be removed from solution either by freeze-drying or by crystallisation.

The scheme outlined in this survey does not necessarily hold for all proteins; often one or more stages are omitted or alternative procedures are carried out.

EXPERIMENT 11
PROPERTIES OF AMINO ACIDS

Reagents

Solid glycine
Solid tyrosine
0·1% w/v glycine solution
0·1% w/v cysteine solution
0·02% w/v phenylalanine solution
0·02% w/v tyrosine solution
0·02% w/v tryptophan solution
0·02% w/v arginine solution
Serum diluted tenfold
2 M NaOH
2 M HCl
2 M HNO_3
2 M lead acetate
conc. HNO_3
conc. NH_4OH
conc. H_2SO_4
40% w/v NaOH
10% w/v $NaNO_2$
0·1 M acetate buffer pH 5
0·2% w/v ninhydrin in 95% ethanol
* Millon's reagent
* Hopkin–Cole reagent
0·02% w/v α-naphthol in 95% ethanol
40% w/v urea
Alkaline sodium hypobromite (1 ml Br_2 in 100 ml M NaOH)

Procedure

SOLUBILITY

1. Note the appearance of glycine and tyrosine. Test their solubilities in water and determine the approximate pH of the solution or suspension with universal indicator paper.
2. If the compound is insoluble or sparingly soluble in water,

determine whether it is more soluble in (a) 2 M HCl and (b) 2 M NaOH. Comment on the results.

3. Isoelectric point precipitation: dissolve 2 spatula-points (about 0·5 g) of tyrosine in 1 ml of water containing 2 drops of 2 M NaOH (warming if necessary) and add 1 drop of methyl red indicator. Add 0·2 M HCl (2 M HCl diluted tenfold) dropwise with constant shaking, until the indicator just changes colour. Repeat the experiment, dissolving the tyrosine in 1 ml of water containing 2 drops of 2 M HCl and 1 drop of methyl red, and using 0·2 M NaOH to adjust the pH. Comment on the results.

REACTION WITH NITROUS ACID

Dissolve 2 spatula-points (about 0·5 g) of glycine and tyrosine in 2 ml of 2 M HCl in two test-tubes. Add 2 ml of 2 M HCl to a third test-tube to act as a control. Cool the tubes in ice and add 1 ml of ice-cold 10% $NaNO_2$ to each tube. Comment on the result.

REACTION WITH NINHYDRIN

Dissolve about a quarter of a spatula-point (about 0·1 g) of glycine in 2 ml of buffer (pH 5) and add 1 ml of ninhydrin solution. Repeat with tyrosine. Heat the solutions in a boiling water-bath for 10 min.

Carry out Experiments 4–8 with the six amino acid solutions and with diluted serum.

XANTHOPROTEIC TEST

This test is positive for aromatic amino acids. It is based on the nitration of the benzene nucleus to yield a yellow derivative which turns orange on addition of ammonia.

To 2 ml of test solution add 2 ml of conc. HNO_3 and heat the mixture in a boiling water-bath for 30 s. Cool, then add dropwise conc. NH_4OH until the solution is alkaline. Observe any colour changes.

MILLON'S TEST

This test is positive for phenolic amino acids, e.g. tyrosine. The reagent is mercuric sulphate in nitric acid. A pink coloration or pink precipitate indicates a positive reaction.

First ensure that the test solution is neutral or acid; if it is not, acidify the solution with 2 M HNO_3 (not with HCl, as Cl^- ions interfere with the test). To 3 ml of the test solution add 2 drops of Millon's reagent. Heat the solution in a boiling water-bath for 5 min and observe any colour changes.

HOPKIN–COLE TEST

This test is positive for amino acids containing the indole ring, e.g. tryptophan. The reagent is glyoxylic acid (CHO—CO_2H) in conc. H_2SO_4. A purple ring at the interface indicates a positive reaction.

To 3 ml of test solution add 1 ml of Hopkin–Cole reagent and mix thoroughly. Carefully pour conc. H_2SO_4 down the side of the tube so as to form two layers. Observe any changes which occur.

SULPHUR TEST

The sulphur of cysteine and cystine is converted to inorganic sulphide by boiling with conc. NaOH. Lead acetate is added and a precipitate of black lead sulphide indicates a positive reaction. (The reaction is negative for methionine, a thioether, as the sulphur is not converted to inorganic sulphide.)

To 3 ml of test solution add 1 ml of 40% NaOH and 1–2 drops of lead acetate solution. Heat in a boiling water-bath for 3 min. Observe any changes which occur.

SAKAGUCHI TEST

This test is positive for compounds containing the guanidine group, e.g. arginine. The reagent is α-naphthol and sodium hypochlorite or hypobromite in an alkaline medium. A red coloration indicates a positive reaction.

To 5 ml of test solution add 1 ml of 2 M NaOH and 1 ml of ethanolic 0·02% α-naphthol. Mix well and cool tube in ice. Add about 1 ml of alkaline hypobromite solution, mix well and after 30 s add 1 ml of urea solution (urea destroys the excess hypobromite). Observe any changes which occur.

EXPERIMENT 12
PROPERTIES OF PROTEINS IN SERUM

Reagents

2 M NaOH
2 M HCl
0·2 M $HgCl_2$
2 M $CuSO_4$
2 M $BaCl_2$
2 M lead acetate
2 M $AgNO_3$
2 M HNO_3
40% w/v NaOH
conc. H_2SO_4
conc. HNO_3
conc. HCl
Ethanol 95% v/v
Solid $(NH_4)_2SO_4$
Saturated $(NH_4)_2SO_4$ solution
0·1 M acetic acid
0·1 M sodium acetate
10% w/v trichloracetic acid
10% w/v picric acid
10% w/v sulphosalicylic acid
10% w/v phosphotungstic acid
1 : 10 diluted serum
Viscose tubing
* Fehling's solution No. 1

Procedure

BIURET TEST

Biuret reagent is dilute cupric sulphate in an alkaline solution. Compounds containing two or more peptide linkages (e.g. biuret, proteins, etc.) complex with cupric ions to give a characteristic purple colour.

$$
\begin{array}{c}
H_2O \\
| \\
H_2N: \quad | \quad :NH_2 \\
\diagdown \quad | \quad \diagup \\
O{=}C \quad \diagdown \quad | \quad \diagup \quad C{=}O \\
HN \quad \quad Cu^{2+} \quad \quad NH \\
O{=}C \quad \diagup \quad | \quad \diagdown \quad C{=}O \\
\diagup \quad | \quad \diagdown \\
H_2N: \quad | \quad :NH_2 \\
| \\
H_2O
\end{array}
$$

Cupric biuret complex

To 3 ml of water in one tube (control) and 3 ml of diluted serum in another, add 3 ml of 2 M NaOH. By means of a pasteur pipette add 2–3 drops of dilute copper sulphate solution (Fehling's solution No. 1 diluted twentyfold). Mix well and note any colour change.

DENATURATION BY HEAT

Label three tubes 1, 2 and 3, and to each tube add 2 ml of diluted serum and 2 drops of methyl red indicator (pK, 5·5). To tube 1 add 2 ml of water; to tube 2 add 2 ml of 0·2 M NaOH; and to tube 3 add 2 ml of 0·2 M HCl (2 M acid and alkali diluted tenfold). The colour in tube 1 should be orange (pH, 5·5); if it is yellow, add diluted HCl dropwise until the orange colour is restored. Heat the tubes in a boiling water-bath for 5 min. Note any changes in the tubes. Adjust pH in tubes 2 and 3 by adding dilute acid and alkali, respectively, until the colour of the indicator is orange as in tube 1. Comment on the results.

DENATURATION BY EXTREMES IN pH

Place about 2 ml of diluted serum in a test-tube, incline the tube and slowly add conc. HNO_3 down the side of the tube, to form a layer below the serum. Note the appearance. Mix the contents of the tube and again note the appearance.

Repeat the experiment using conc. H_2SO_4, conc. HCl and 40% NaOH.

ISOELECTRIC POINT PRECIPITATION OF SERUM PROTEINS BY ETHANOL

Make up a series of eight tubes in ml as follows:

Tube	1	2	3	4	5	6	7	8
0·1M sodium acetate	1·0	1·0	1·0	1·0	1·0	1·0	1·0	1·0
0·1M acetic acid	4·0	2·0	1·0	0·0	0·0	0·0	0·0	0·0
0·01M acetic acid	0·0	0·0	0·0	5·0	2·5	1·25	0·6	0·3
Water	2·0	4·0	5·0	1·0	3·5	4·75	5·4	5·7
pH	4·1	4·4	4·7	5·1	5·4	5·7	6·0	6·3

To each tube add 1 ml of diluted serum, mix well and note any small difference in turbidity. Finally, add 5 ml of ethanol to each tube, mix well and allow the rack of tubes to stand undisturbed. Note any differences in turbidity immediately after adding ethanol and at 15-min intervals until 45 min have elapsed.

Tabulate results and comment on approximate isoelectric point of serum proteins.

PRECIPITATION OF SERUM PROTEINS BY HALF-SATURATED AND SATURATED AMMONIUM SULPHATE

To 5 ml of serum (undiluted) in a boiling tube, add 5 ml of water and 10 ml of saturated ammonium sulphate. Mix the contents of the tube and allow it to stand for about 3 min. Filter the mixture into another boiling tube, refiltering the filtrate if any cloudiness remains.

Dissolve a small portion of the precipitate in about 2 ml of water and boil the solution in a boiling water-bath for about 5 min. Also boil an aliquot of the filtrate.

To the remainder of the filtrate add solid ammonium sulphate, thoroughly mixing the contents of the tube and adding more solid ammonium sulphate until the mixture is saturated with the salt. Allow the mixture to stand for a few minutes and then filter. Boil a small aliquot of the filtrate.

Comment on the results and state the fractions in which albumen and the globulin proteins appear.

PRECIPITATION OF PROTEINS BY COMPLEX ANIONS

Prepare four tubes containing 2–3 ml of diluted serum. To the first add 10% trichloracetic acid dropwise until an excess has been added. Adjust the pH with 2 M NaOH to approximately 10, using universal indicator paper. Repeat the experiment using 10% sulphosalicylic acid, 10% picric acid and 10% phosphotungstic acid. Note whether the precipitates are soluble in excess of the reagent or in alkaline conditions.

PRECIPITATION OF PROTEINS BY CATIONS OF HEAVY METALS

Prepare four tubes containing 2–3 ml of diluted serum. To the first tube add mercuric chloride dropwise until an excess has been added. Note results. Then adjust the pH, using universal indicator paper, to about 10 with dilute NaOH. Mix the contents of the tube and note any change. Finally, add dilute acid until the pH is about 2. Repeat the experiment using dilute solutions of barium chloride, cupric sulphate and lead acetate. Comment on the results.

DIALYSIS

Fill a viscose tube sealed at one end with dilute serum. Seal the other end of the tubing and suspend it in about 200 ml of distilled water in a beaker. Continuously stir the dialysate with a magnetic stirrer. After about 30 min, test aliquots of the dialysate and aliquots of the dialysis mixture for Cl^- ions ($AgNO_3$+dil. HNO_3), SO_4^{2-} ions ($BaCl_2$+dil. HCl) and protein (biuret test).

Dialysis will be incomplete after this time and therefore the dialysis mixture will still contain crystalloids. Replace the dialysate with fresh distilled water and continue dialysis overnight. Re-test the dialysate and dialysis mixture for Cl^- ions, SO_4^{2-} ions and protein.

EXPERIMENT 13
SEPARATION OF AMINO ACIDS BY
ION EXCHANGE CHROMATOGRAPHY

A sulphonic acid cation exchange resin (manufactured under the trade names Dowex 50 and Zeo-Karb 225) is used in this experiment for the separation of amino acids.

Stein and Moore (1951) separated quantitatively all the amino acids of a protein hydrolysate, by fractional elution from a Dowex 50 column. Amino acids at pH's lower than their respective isoelectric points exist as cations, and if the pH of the hydrolysate is adjusted to below 3, complete uptake of all the amino acids occurs. The amino acids are selectively eluted from the column by a gradient buffer of increasing pH. Their order of appearance is mainly due to their dissociation constants and, to a lesser extent, their size. Temperature also affects the rate of their elution.

In this experiment three amino acids, one acidic, one neutral and one basic, are separated on a Zeo-Karb 225 column in the Na^+ form. As these amino acids have very different dissociation constants, elution can be carried out with two solutions of different pH values.

Reagents

50 mM citrate buffer at pH 4·2
0·1 M NaOH
Zeo-Karb 225 (52–100 mesh; 4·5% ×-linking agent; Na^+ form)
2 M HCl
2 M NaOH
Neutral ninhydrin solution   (0·2 g in 100 ml acetone)
Acidic ninhydrin solution   (0·2 g in 90 ml acetone, 5 ml glacial HAc and 5 ml H_2O)
Amino acid solution of: Add 1 ml 1% w/v aspartic acid solu-
   DL-aspartic acid   tion, 1 ml 1% w/v alanine solution
   DL-alanine   and 1 ml 1% w/v lysine solution in
   DL-lysine   0·1 M HCl to 2 ml citrate buffer, and mix well
0·1 M acetate buffer at pH 5

Procedure

PREPARATION OF COLUMN

Place about 10 g of resin in a 100-ml beaker, add 25 ml of 2 M HCl and stir well. Decant off the acid and rinse the resin three times with distilled water. Add 25 ml of 2 M NaOH to the resin, stir well, and again decant off the liquid. Thoroughly rinse the resin with distilled water until the washings are neutral to indicator paper. Suspend the resin in about 50 ml of citrate buffer at pH 4·2.

Take a 25-ml burette with a coarse tip (if necessary shorten the tip) and plug the end of the burette with a small piece of glass wool. Clamp the burette in position and pour the slurry of resin into the column. Stir the resin with a glass rod; this ensures that the column is evenly packed and contains no air-bubbles. Allow the resin to settle and run off the excess liquid. NEVER, at any time, allow the column to run dry. Add more slurry, stirring to remove air-bubbles, until the resin has settled to a height of about 8 cm. Gently pipette more buffer down the side of the burette and slowly allow buffer through the column until the level of the liquid is about 1 cm above the level of the resin. In order to prevent disturbance of the resin surface, place on it a small disc of filter paper, and introduce the solutions to the column by pipetting them down the side of the burette.

Pass citrate buffer through the column until the effluent is pH 4·2. Allow excess buffer through the column until the liquid surface is just above the filter paper disc.

APPLICATION OF SAMPLE AND COLLECTION OF FRACTIONS

Carefully pipette 0·1 ml of the amino acid solution on to the column and allow sample to flow into the resin. Wash through with 1 ml of citrate buffer and then with a further 2 ml of citrate buffer.

Pour 50 ml of citrate buffer at pH 4·2 into a separating funnel to serve as a reservoir. The outlet of the reservoir is connected to a pasteur pipette by a short piece of polythene tubing. The pasteur pipette is supported by a rubber bung which has an air-escape. The apparatus is shown in Figure 3.11.

Number a series of test-tubes, 1–25, for collecting fractions.

Allow the buffer to flow through the column at a rate of 2 ml/3 min. Collect 2-ml fractions in the series of numbered test-tubes.

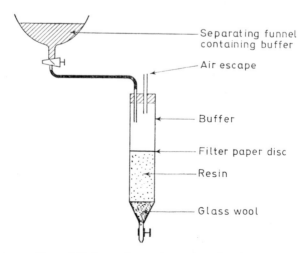

Figure 3.11. Ion exchange chromatography apparatus

The elution of amino acids is determined by the ninhydrin test. When each group of four successive fractions has been collectedt test for amino acids as follows: to 1 ml of fraction add 1 ml of buffer at pH 5 and 1 ml of neutral ninhydrin solution, and heat in a boiling water-bath for 10 min. A blue coloration indicates a positive reaction.

After the second amino acid has been eluted with pH 4·2 buffer, elute two ninhydrin-negative fractions. The total number of fractions collected with this buffer should be in the region of 13–14. Close the reservoir and withdraw excess buffer at pH 4·2 with a pasteur pipette (taking care not to disturb the surface of the resin).

Pipette 2 ml of 0·1 M NaOH onto the column. Replace the buffer in the reservoir with 0·1 M NaOH and collect 2-ml fractions as before. When testing the fractions for amino acids, use the acidic ninhydrin solution. After the third and last amino acid has been eluted with 0·1 M NaOH, collect two ninhydrin-negative fractions. The total number of fractions collected with dilute alkali should be in the region of 7–10.

Read absorbance given by the ninhydrin reaction in the series of tubes against a water blank at 570 nm (yellow filter).

Plot absorbance against the fraction number of the sample. This plot is called an elution curve.

EXPERIMENT 14
SEPARATION OF AMINO ACIDS BY ELECTROPHORESIS

Electrophoresis is the separation of charged or partially charged molecules by an applied electric current. In moving boundary electrophoresis the charged particles migrate through a solution, and in zone electrophoresis the charged particles migrate along a porous medium such as paper, cellulose acetate, starch gel, agar or polyacrilamide. In this experiment we are using zone electrophoresis.

The two ends of a strip of the porous medium are suspended in a buffer solution and the strip is moistened with the buffer to allow the flow of current. The degree of migration is dependent upon the amount of charge present in the molecules; the greater the charge, the greater the migration. Positively charged molecules move towards the cathode and negatively charged molecules towards the anode. Electrophoresis may be considered as incomplete electrolysis.

The conditions throughout the experiment do not remain constant. As the experiment proceeds, heat is generated which lowers the resistance of the strip and thus the mobility of the molecules increases with time. This effect is slight with low-voltage electrophoresis. If the variables are constant from one experiment to another, results from individual experiments can be compared.

Colourless substances are located with a suitable locating agent at the end of electrophoresis.

In paper electrophoresis large molecules such as proteins are retarded by the paper owing to adsorption; this also leads to streaking and thus paper is a poor medium for separation of charged macromolecules.

The simplest electrophoresis apparatus, as shown in Figure 3.12, consists of two beakers containing electrodes immersed in buffer (platinum wire or carbon rods are suitable electrodes), with a strip of filter paper, with its ends dipping into the buffer, in each beaker. The strip is sandwiched between two glass plate supports and the electrodes are connected to a high-voltage battery or power pack.

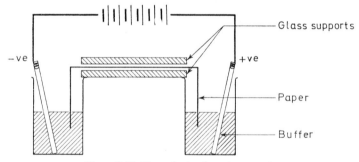

Figure 3.12. Zone electrophoresis apparatus

Commercially available apparatus comes in two types: the vertical type, where the strip is suspended by a glass rod with its ends hanging down, and the horizontal type, where the strip is kept horizontal.

In this experiment three amino acids—one neutral, one acidic and one basic—are separated by paper electrophoresis at different pH values. Amino acids do not migrate at their isoelectric points; at pH values less than the isoelectric point, amino acids possess a positive charge and move towards the cathode; at pH values greater than the isoelectric point, amino acids possess a negative charge and move towards the anode.

This experiment should be shared between the class, each student performing the electrophoresis at a different pH value.

Reagents and Apparatus

Buffers at pH 1·9 58 ml glacial acetic acid + 26 ml 25% w/v formic acid in 2 litres
                4·0 10·2 g potassium hydrogen phthalate per litre
                6·0 phosphate buffer
                8·0 tris (hydroxymethyl) aminomethane/HCl buffer
              10·0 boric acid/NaOH buffer
              11·5 5·3 g anhydrous Na_2CO_3 per litre

Neutral ninhydrin solution (0·2 g in 100 ml acetone)
Acidic ninhydrin solution (0·2 g in 90 ml acetone + 5 ml glacial acetic acid and 5 ml H_2O)
Alkaline ninhydrin solution (0·2 g in 100 ml ethanol + 0·5 ml M KOH)
Aspartic acid solution ($\simeq 0.1\%$ w/v)
Alanine solution   ($\simeq 0.1\%$ w/v)
Lysine solution   ($\simeq 0.1\%$ w/v)
Mixture of the three amino acids (equal volume mixture)
Electrophoresis apparatus
Power pack (Shandon)   Four tanks can be coupled to the same
or   unit at once
HV battery (120 V)

Procedure

Place an equal volume of buffer at pH 1·9 in the two buffer compartments.

Cut a strip of paper from Whatman No. 1 filter paper (cut strip along the grain of the paper) to fit the apparatus. Draw a line across the centre of the strip and mark four origins, evenly spaced, along this line. Mark the anode and cathode at the ends of the paper.

Place buffer at pH 1·9 in a dipping-tray and dip the strip through the buffer, then thoroughly blot the strip between sheets of filter paper. Place the strip in position in the apparatus. On origin 1 spot, by means of a capillary pipette or platinum loop, one drop of aspartic acid; on origin 2 spot one drop of alanine; on origin 3 spot one drop of lysine; and on origin 4 spot one drop of the mixture. Connect the electrodes to the power pack (+ve to +ve, −ve to −ve) at 200–250 V and allow current to run for at least 4 h.

Disconnect the current; remove and dry the strip in a current of hot air. Place alkaline ninhydrin solution in a dipping-tray, dip the strip and then heat at 105°C to develop the spots.

Repeat the experiment at pH 4, 6, 8, 10 and 11·5. At pH 4, 6 and 8 dip the strip in neutral ninhydrin solution, and at pH 10 and 11·5 dip the strip in acidic ninhydrin solution.

Tabulate the mobilities of the three amino acids at different pH values and deduce their approximate isoelectric points.

EXPERIMENT 15
POTENTIOMETRIC TITRATION OF AMINO ACIDS

Determination of pK_a, pK_b and Isoelectric Point of Glycine

Amino acids contain an acidic and a basic group in the same molecule; for this reason they act as efficient buffers at certain pH values.

Consider the equation:

$$\underset{\text{Base}}{\overset{NH_2}{\underset{CO_2^-}{\overset{|}{C}H_2}}} \underset{H^+}{\rightleftarrows} \underset{\text{Salt}}{\overset{\overset{+}{N}H_3}{\underset{CO_2^-}{\overset{|}{C}H_2}}} \underset{H^+}{\rightleftarrows} \underset{\text{Acid}}{\overset{\overset{+}{N}H_3}{\underset{CO_2H}{\overset{|}{C}H_2}}}$$

The isoelectric point is the pH at which the amino acid carries no net charge, i.e. the salt form; it is in this form that the amino acid possesses no buffering action. The buffering action is greatest when the concentration of the salt form equals the concentration of the acid or base form.

Consider the reaction acid ⇌ salt:

$$K_a = \frac{[\text{salt}][H^+]}{[\text{acid}]}$$

where K_a = equilibrium constant;

$$[H^+] = K_a \cdot \frac{[\text{acid}]}{[\text{salt}]}$$

$$pH = pK_a + \log_{10} \frac{[\text{salt}]}{[\text{acid}]}$$

where $pK_a = -\log_{10} K_a$;

$$pK_a = pH$$

when [salt] = [acid]. This is in the acid pH range of greatest buffering capacity.

Similarly for the reaction salt ⇌ base:

$$pH = pK_b + \log_{10}\frac{[base]}{[salt]}$$

and

$$pK_b = pH$$

when [base] = [salt]. This is in the alkaline pH range of greatest buffering capacity.

At the isoelectric point the concentration of acid equals the concentration of base:

$$K_a = \frac{[H^+][salt]}{[acid]}; \quad [acid] = \frac{[H^+][salt]}{K_a}$$

$$K_b = \frac{[H^+][base]}{[salt]}; \quad [base] = K_b \cdot \frac{[salt]}{[H^+]}$$

At the isoelectric point (IEP)

$$[acid] = [base]$$

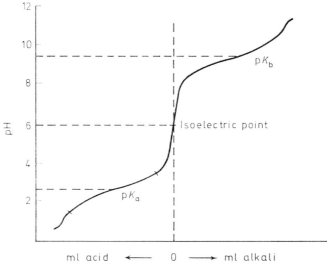

Figure 3.13. *A typical titration curve of a monoamino, monocarboxylic amino acid*

Therefore

$$\frac{[H^+][\text{salt}]}{K_a} = K_b \cdot \frac{[\text{salt}]}{[H^+]}$$

$$[H^+]^2 = K_a \cdot K_b$$

$$pH_{IEP} = \frac{pK_a + pK_b}{2}$$

It will be seen that the titration curves of monoamino monocarboxylic amino acids are similar. A typical curve is shown in Figure 3.13.

Reagents and Apparatus

0·25 M H_2SO_4
0·5 M NaOH
Glycine (Analar)
Buffer tablets
pH meter

Procedure

Standardise the pH meter against a pH 6·99 buffer made from a buffer tablet. Prepare 100 ml of 0·1 M glycine solution (0·75 g/100 ml). Use 20 ml of this solution for titration with acid and 20 ml for titration with alkali. Place the glass and reference electrodes in the glycine solution, adding distilled water if necessary to cover the tips of the electrodes. Record the pH. Titrate the glycine solution against 0·25 M H_2SO_4. Add one drop at a time for the first six drops and then slowly increase the increments of acid until the pH is $\simeq$ 1·5. Stir the titration mixture continuously by means of a magnetic stirrer. Repeat the procedure, titrating glycine against 0·5 M NaOH. Add the caustic soda in drops at first, and then increase the increments until the pH is $\simeq$ 11.

Tabulate volume of acid or alkali for each pH value. Plot pH against volume of acid or alkali. Determine the isoelectric point and pK values of glycine from the graph.

Compare your pK values and the isoelectric point with the accepted values.

EXPERIMENT 16
ESTIMATION OF AMINO ACIDS
(THE FORMOL TITRATION)

Amino acids contain an amino and a carboxylic group which can be titrated against an acid or alkali, respectively, but the pH of the neutralisation points lies in the region of 2 and 12 (owing to the hydrolysis of the salts formed) and there are no useful indicators for these ranges.

The amino group can be shielded by the addition of formaldehyde and the pH of the neutralisation point of the resultant acid is shifted from 12 to 9:

$$R-CH \begin{matrix} CO_2H \\ NH_2 \end{matrix} + H.CHO \longrightarrow R-CH \begin{matrix} CO_2H \\ N-CH_2OH \\ | \\ H \end{matrix}$$

Amino acid formaldehyde complex (resultant acid)

At this pH, phenolphthalein can be used as an indicator.

Reagents

40% formaldehyde
Standardised sodium hydroxide solution (approx. 0·1 M)
Named amino acid of 'unknown strength' (0·07–0·13 M)
Phenolphthalein

Procedure

The formaldehyde must first be neutralised as it is likely to contain small amounts of free acid. For each titration, mix 10 ml of 40% formaldehyde with 20 ml of water and add four drops of phenolphthalein. Add standardised NaOH carefully from a burette until the faintest permanent pink colour is produced. It is unnecessary to record the amount of NaOH added. Neutralise enough formaldehyde for three titrations. (40% formaldehyde is highly corrosive and must not be pipetted.)

Pipette 25 ml of the amino acid solution into a conical flask, add four drops of phenolphthalein and neutralise the solution with

NaOH, adding the alkali dropwise, until the faintest permanent pink colour is produced. It is again unnecessary to record the amount of alkali added.

Add 30 ml of the neutralised formaldehyde solution to the amino acid solution (note the disappearance of the pink colour), and titrate the contents of the flask against the standard NaOH until a pink colour is produced as before. Record the titre value. Carry out the titration in triplicate.

Calculation

Let

$$\text{Molarity of NaOH} = y \text{ M}$$
$$\text{titre value} = b \text{ ml}$$

1 mol NaOH ≡ 1 mol amino acid ≡ 14 g amino nitrogen

b ml (titre) $\times y$ M NaOH ≡ $b \times y$ mmol amino acid

≡ $14 \times b \times y$ mg N/25 ml amino acid solution

Express the results as: (1) g/l amino nitrogen, (2) g/l amino acid.

EXPERIMENT 17
ISOLATION OF CASEIN

Casein is insoluble at its isoelectric point and therefore the protein is readily precipitated by adjusting the pH of the solution.

Reagents

Milk
10% acetic acid
1 M sodium acetate
95% v/v ethanol
Ether

Procedure

Place 30 ml of milk and 30 ml of water in a 250-ml beaker and warm the mixture to about 40°C. Add dropwise, with constant stirring, about 2 ml of 10% acetic acid. Allow the mixture to stand for 5 min and then add 2 ml of 1 M sodium acetate (this buffers the mixture to ≃ pH 4·5). Stir, cool to room temperature or below and allow mixture to stand for a further 5 min. Filter the mixture through muslin (or centrifuge). Wash the precipitate with a small amount of water and re-filter through muslin. Suspend the precipitate in about 10 ml of 95% alcohol and then filter at the pump. Rinse precipitate with alcohol, then re-suspend the precipitate in about 10 ml of ether and filter at the pump. Rinse the precipitate in ether and finally remove all traces of solvent by pressing the casein between sheets of filter paper.

Weigh the dry casein and determine the percentage yield. (Take the density of milk to be 1·0 g/cm^3 and the theoretical casein content of milk to be 3·5 g %.)

EXPERIMENT 18
SEPARATION OF AMINO ACIDS OF A PROTEIN HYDROLYSATE BY PAPER CHROMATOGRAPHY

Both acids and alkalis hydrolyse the peptide linkages of proteins, but during hydrolysis degradation of some of the constituent amino acids occurs. Hydrolysis with acids, usually 6 M HCl, is preferred as acids destroy fewer of the amino acids.

Amino acids of protein hydrolysates can be separated by paper chromatography. No one solvent or solvent mixture is capable of separating all the amino acids, but complete separation can be achieved by two-dimensional chromatography.

An aliquot of the protein hydrolysate is spotted at one corner of the chromatogram and the amino acids are partially separated by running in the first solvent. The chromatogram is dried and turned through 90° so that the partially separated acids lie along the origin, and it is then run in the second solvent (see Figure 3.14).

Figure 3.14(b) represents partially separated amino acids after the chromatogram has been run in the first direction. (The chromatogram is usually run in S_1 and S_2 over two consecutive nights.)

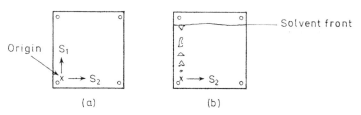

Figure 3.14. Two-dimensional paper chromatography

Reagents and Apparatus

6 M HCl
20% ethanol
0·2% w/v ninhydrin in acetone
Casein (from previous experiment)
Solvent 1: BuOH/HAc/H_2O 120/30/50 by volume
Solvent 2: PhOH/H_2O/NH_3 160/40/1 by volume
Square chromatography tank (to fit 30 cm × 30 cm chromatograms)
Chromatography frame
Micro-syringe or capillary pipette

Procedure

Place about 50 mg of casein in a 250-ml round-bottomed flask. Add 50 ml of 6 M HCl, attach a vertical water-jacketed condenser and reflux the mixture for 20 h. Rearrange the apparatus for distillation and distil off the HCl to dryness. Dissolve the residue in about 5 ml of warm water, filter and evaporate the aqueous solution to dryness. Dissolve the residue in 2 ml of 20% ethanol and transfer the ethanolic solution of amino acid hydrochlorides to a labelled Quickfit test-tube.

Place 150–200 ml of solvent 1 in the chromatography tank and allow the tank to equilibrate.

Prepare 30 cm × 30 cm chromatography paper (the paper should have a hole at each corner so that it fits into the frame) by marking the origin 3 cm from one corner towards the centre of the paper. Draw two lines at right angles to each other from the origin and indicate the direction of solvent 1 and solvent 2.

Spot, by means of a micro-syringe, 10 μl of amino acid hydrochloride solution, allowing the solution to dry between addition of each drop. Place the chromatogram in the frame and run in solvent 1 overnight ($\simeq$ 16 h). Remove the chromatogram from the tank, mark the solvent front and allow the paper to dry in a current of cold air (about $\frac{1}{2}$ h).

Thoroughly rinse the tank and prepare solvent 2. This solvent is usually prepared in bulk by adding 125 ml of water to 500 g of phenol in the original dark bottle, replacing the stopper and immersing the bottle in a water-bath at about 40°C until a homogeneous solution is obtained. The requisite amount of NH_3 is added to the phenol solution just before the solvent is used. (Care must be taken not to handle phenol as it is highly corrosive.) Place 150–200 ml of solvent 2 in the tank and allow the tank to equilibrate. Turn the chromatogram, which is still in the frame, through 90° and place in solvent 2. Allow the chromatogram to run overnight ($\simeq$ 16 h). Remove from the tank, mark the solvent front and allow the paper to dry in a current of cold air (about 4 h).

Place 0·2% ninhydrin in the dipping tray and dip the chromatogram. Heat at 105°C for 10 min to develop the spots.

Outline the individual spots and compare the amino acid pattern with a standard amino acid map (I. SMITH (Ed.), *Chromatographic and Electrophoretic Techniques*, Vol. II, Heinemann, London, 103 (1960)).

EXPERIMENT 19
ULTRA-VIOLET SPECTRA OF TYROSINE

Proteins absorb in the ultra-violet region of the electromagnetic spectrum. This characteristic can be used to estimate proteins quantitatively in solution. The UV absorption is due to aromatic amino acids, namely tyrosine, tryptophan and, to a lesser extent, phenylalanine.

Reagent and Apparatus

0·001 M tyrosine
UV spectrophotometer
Silica cells

Procedure

1. Switch on UV spectrophotometer.
2. Adjust lamp to deuterium source.
3. Switch on extra power-pack.
4. Place (a) water and (b) 0·001 M tyrosine solution in 1-cm silica cuvettes.

Read the absorbance of the tyrosine solution in the range 240–320 nm at 5-nm intervals, adjusting the absorbance against water to zero at each wavelength. Return and confirm the position of the peak by taking readings at 2-nm intervals around the maximum.

Plot a graph of the UV spectra, by plotting absorbance against wavelength, and note the wavelength of maximum absorption.

EXPERIMENT 20
QUANTITATIVE ESTIMATION OF PROTEINS

The Biuret Method

This test is reliable as a method of determining protein concentration but requires relatively large quantities of protein (range 1–20 mg).

The Folin–Ciocaltea method

This method is more sensitive (range 25–500 µg). Colour formation is due to the formation of a cupric complex with an alkaline copper reagent as in the biuret test and to the reduction of phosphomolybdates and phosphotungstates in the reagent by phenolic compounds, e.g. tyrosine. This method therefore determines the amount of tyrosine present in proteins. For accurate determinations, the protein of unknown concentration should be compared with the same protein of known concentration so that the percentage of tyrosine in the two protein solutions is constant.

Reagents and Apparatus

    Standard protein solution containing 20 mg/2 ml
* Biuret reagent
    Serum diluted tenfold
    Solution A: 2% w/v Na_2CO_3 (anhydrous) in 0·1 M NaOH
    Solution B: 0·5% w/v $CuSO_4.5H_2O$ in 1% sodium or potassium tartrate
    Solution C: mix 50 ml solution A with 1 ml solution B just before use
* Folin–Ciocaltea reagent
    Colorimeter or spectrophotometer

Procedure

BIURET TEST

Pipette, in duplicate, 0, 0·4, 0·8, 1·2, 1·6 and 2·0 ml of standard protein solution into a series of test-tubes. Make volumes up to 2 ml with water. In addition, prepare two tubes containing 2 ml of diluted serum. Add 8 ml of biuret reagent to each tube and mix well. After 30 min read absorbance against reagent blank at 570 nm (yellow filter).

Plot a standard protein curve and determine the protein concentration of serum in g %.

FOLIN–CIOCALTEAU TEST

Dilute the standard protein solution twentyfold (pipette 2·5 ml into a 50-ml volumetric flask and make up to the mark). Pipette, in duplicate, 0, 0·05, 0·1, 0·25, 0·5 and 1·0 ml of the diluted standard protein solution into a series of test-tubes. Make volumes up to 1 ml with water. In addition, prepare two tubes containing 1 ml of diluted serum, diluted a further twentyfold. Add 5 ml of solution C to each tube, mix well and allow the tubes to stand for 10 min at room temperature. Add 0·5 ml of Folin–Ciocaltea reagent to each tube with immediate mixing. After 30 min, read absorbance against reagent blank at 750 nm (deep red filter).

Plot a standard protein curve and determine the protein concentration of serum in g %.

EXPERIMENT 21
KJELDAHL'S ESTIMATION OF NITROGEN

All proteins contain nitrogen, the average content being 16%. Nitrogen determinations are therefore often carried out to estimate proteins quantitatively. Kjeldahl's estimation determines total nitrogen and there are many naturally occurring substances which also contain nitrogen, e.g. nucleic acids, certain lipids, creatinine, urea, etc. Therefore protein nitrogen is determined by the difference between total nitrogen and non-protein nitrogen. Non-protein nitrogen is determined by precipitating all the proteins with a protein-precipitating agent and determining the nitrogen content of the filtrate.

Organic nitrogen-containing compounds are digested with concentrated sulphuric acid in the presence of a catalyst, such as selenium dioxide, copper sulphate or potassium sulphate, to yield inorganic ammonium sulphate. An excess of concentrated caustic soda is added to the digested mixture and the liberated ammonia is steam-distilled into saturated boric acid. The amount of ammonium borate formed is determined by titration with a standard acid, using a suitable indicator. Alternatively, the liber-rated ammonia can be steam-distilled into a known amount of standard acid, and the amount of acid used in neutralising the ammonia can be determined by back-titration.

Reagents and Apparatus

* Acid digestion mixture
  Saturated boric acid
  40% w/v NaOH
  Ammonium sulphate solution of 'unknown strength' (0·01–0·07 M)
  Standardised HCl (approx. 0·025 M)
  Serum
  Indicator: 1 part methyl red, 3 parts bromcresol green
  Digestion rack
  Kjeldahl flasks
  Markham still
  Steam generator

Procedure

DETERMINATION OF NITROGEN CONTENT OF A GIVEN AMMONIUM SULPHATE SOLUTION

Set up the apparatus as in Figure 3.15.

Heat the water in the steam generator. Allow steam to pass through the apparatus for about 10 min. Empty the distillation

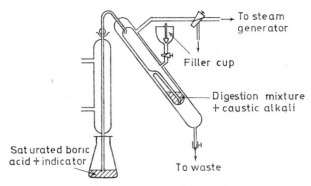

Figure 3.15. A Markham still

flask. (This is done by removing the heat, which causes the liquid in the flask to be sucked back and so to be removed via the lower cock.) Allow steam to pass gently through the apparatus.

Fill the conical flask with sufficient saturated boric acid to cover the tip of the condenser and add a few drops of the mixed indicator. Pipette 5 ml of 'unknown strength' ammonium sulphate solution into the filler cup and run this solution slowly into the distillation flask. Add 5 ml of 40% NaOH into the cup and allow it, in small amounts at a time, into the flask. Thoroughly wash the cup with water (to prevent the seizing up of the ground-glass joint) and allow washings into the flask. Steam-distil the mixture for about 5 min or until about 25 ml of distillate has been collected. Lower the receiver and wash the tip of the condenser into the distillate. Titrate the distillate against standard HCl. Repeat the titration three times and carry out a titration blank.

Calculation

Let

$$\text{molarity of HCl} = y \text{ M}$$
$$\text{titre value} = b \text{ ml}$$
$$1 \text{ mol HCl} \equiv 1 \text{ mol ammonia} \equiv 14 \text{ g nitrogen}$$
$$b \text{ ml (titre)} \times y \text{ M HCl} \equiv b \times y \text{ mmol ammonia}$$
$$\equiv 14 \times b \times y \text{ mg N/5 ml ammonium sulphate solution}$$

Express results as g N/l ammonium sulphate solution.

DETERMINATION OF TOTAL NITROGEN CONTENT OF SERUM

Pipette, in duplicate, 0·25 ml of serum into a micro-Kjeldahl flask and add, using a safety pipette, 2 ml of acid digestion mixture and a small amount of porous pot to prevent bumping. In addition, carry out a blank containing 2 ml of acid digestion mixture. Gently heat the flask on the digestion rack for 5 min and then increase the heat until the mixture boils more vigorously. Continue digestion for 2–4 h after the mixture has turned colourless. (Carry out digestion in a fume-cupboard if the apparatus does not have a hood connected to a pump to draw off the fumes.)

Cool the digested mixture and add slowly, with constant shaking, about 5 ml of water. Transfer the mixture quantitatively to the filler cup of the Markham still (which has been previously rinsed with steam) and allow it into the distillation flask. Rinse the cup with water and allow the washings into the flask. Add 10 ml of 40% NaOH into the flask via the cup and continue the assay for nitrogen as above.

Repeat the assay and also determine the nitrogen content of the reagent blank.

Calculate the total nitrogen content of the serum and express the results as g N/100 ml serum.

(Non-protein nitrogen accounts for about 3% of the total nitrogen content of serum.)

EXPERIMENT 22
NESSLER'S ESTIMATION OF NITROGEN

Nessler's reagent (potassium mercuric iodide in alkali) is used for colorimetric estimation of inorganic nitrogen (range 5–100 μg N).

Organic nitrogenous compounds are digested to inorganic ammonium sulphate with conc. H_2SO_4 plus catalyst, as in Kjeldahl's estimation. Nessler's reagent is added to a suitable aliquot of the digested mixture; this reacts with the ammonium salt to form a coloured colloidal sol. A small amount of gum ghatti may be added to stabilise this sol.

Reagents and Apparatus

* Nessler's reagent
Ammonium sulphate (Analar)
* Gum ghatti solution
* Acid digestion mixture
40% w/v NaOH
Casein (from Experiment 17)
Colorimeter or spectrophotometer
Micro-Kjeldahl flasks
Digestion rack

Procedure

1. STANDARD NITROGEN CURVE

Prepare a standard solution of ammonium sulphate containing 100 μg N/ml. Pipette, in duplicate, 0·1, 0·2, 0·3, 0·4, 0·5, 0·75 and

1·0 ml of the standard ammonium sulphate solution into a series of test-tubes. Make the volumes up to 5 ml with distilled water. Also prepare a blank containing 5 ml of water. Add 0·5 ml of Nessler's reagent to each tube followed by 0·2 ml of gum ghatti. Mix the contents thoroughly. Allow the tubes to stand for 15 min for maximum colour development.

Read absorbance against the reagent blank at 480 nm (blue filter) and plot a standard nitrogen curve.

2. PERCENTAGE NITROGEN IN CASEIN SAMPLE

Weigh accurately 30–50 mg of your casein sample and quantitatively transfer it to a Kjeldahl digestion flask. Digest the nitrogen content of casein to inorganic ammonnium sulphate as in Experiment 21. Also, prepare a digestion blank.

Allow the digested mixture to cool and add, with care, about 5 ml of water. Adjust the pH to 5 using universal indicator paper, by careful dropwise addition of 40% NaOH. Transfer the mixture quantitatively to a 100-ml volumetric flask and make up to the mark. Repeat this procedure with the reagent blank.

Pipette, in duplicate, 1 ml of digested casein and 1 ml of reagent blank into a series of test-tubes. Make volumes up to 5 ml with water and continue the assay for nitrogen as above. Determine the concentration of the unknown from the standard curve.

Express the result as: (a) g N/100 g casein, and (b) % purity of casein sample (assume that the nitrogen content of casein is 16%).

EXPERIMENT 23
GEL FILTRATION OF DIFFERENT FORMS OF HAEMOGLOBIN

The function of haemoglobin is to transport oxygen to and carbon dioxide from the tissues.

Haemoglobin contains the prosthetic group haem. Haem contains a ferrous ion which is chelated at the centre of a porphyrin ring. Fe^{2+} ions are most stable when a co-ordination number of 6 is satisfied. Four co-ordination positions are filled by the nitrogen atoms of the porphyrin ring and a further position by the base amino acid histidine from a polypeptide chain of the protein; and in oxyhaemoglobin the sixth position is filled by molecular oxygen:

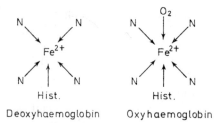

In this experiment three forms of haemoglobin are observed, namely:

1. Methaemoglobin   In this form the Fe^{2+} ion at the centre of the porphyrin ring is oxidised to the ferric form with ferricyanide. This form does not transport O_2 and it is readily observable by its brown colour.
2. Deoxyhaemoglobin This form is purple.
3. Oxyhaemoglobin   Oxygenated haemoglobin—this form is scarlet.

The three forms can be interconverted:

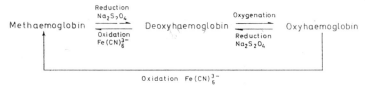

Gel Filtration

Gel filtration is a method of separating substances according to their molecular size. It also provides a convenient method of treating a protein with a reagent; this technique is demonstrated in the following experiment. Methaemoglobin is treated with dithionite on the column; the resultant deoxyhaemoglobin is oxygenated as it passes through the column and oxyhaemoglobin is eluted from the column.

In this experiment a highly cross-linked dextran (Sephadex G25) is used as the gel filtration medium. This medium retains only small molecules (molecular weight of 700 or less) and excludes all other molecules.

Reagents

Sephadex G25 slurry   The gel has been allowed to stand in phosphate buffer (ph 7) for at least 2 h prior to the experiment. This ensures that the particles are fully swollen.

20 mM phosphate buffer
　at pH 7
Potassium ferricyanide
Sodium dithionite　　　Freshly prepared solution of 10 mg/ml
　　　　　　　　　　　　in buffer (ph 7)
Whole blood

Procedure

Place a small piece of glass wool at the end of a 25-ml burette. Evenly pack the column with a Sephadex G25 slurry to form a bed of about 8 cm in height. Care should be taken to use a slurry dilute enough not to trap air in the column while pouring. Wash the column with one column volume of buffer. Ensure that the top of the column is level; if necessary, stir the column to make it level. Allow the excess buffer through the column until its level is just above the bed surface.

Carefully pipette 0·2 ml of a freshly prepared solution of sodium dithionite (10 mg/ml in buffer) onto the surface of the column without disturbing the top of the bed. Allow dithionite to flow into the column stopping just before the meniscus reaches the bed surface, and then carefully pipette 0·2 ml of buffer onto the column surface.

Dilute 1 ml of whole blood tenfold with water, add about 50 mg of solid potassium ferricyanide to the haemolysed cells and mix. Pipette carefully 0·5 ml of the methaemoglobin onto the column and allow the sample to flow into the gel particles. Wash the top of the column, by pipetting about 1 ml of buffer, and then fill the column with buffer (without disturbing the bed) and continue elution. (Do not pause until the protein has been eluted through the dithionite zone, as peroxides may be present in the dithionite and these destroy the haemoglobin.)

Elute the oxyhaemoglobin from the column in a minimum volume of buffer (without eluting any dithionite).

Note the three forms of haemoglobin and the yellow band of ferricyanide which is formed.

CHAPTER 4

Lipids

STRUCTURE AND PROPERTIES OF LIPIDS

Lipids are a group of naturally occurring compounds of great structural diversity. They include simple lipids, waxes, complex lipids, sterols, terpenes, etc. Most lipids are insoluble in water and soluble in organic solvents, and they are essential to living matter.

Simple Lipids

Simple lipids (often referred to as fats, if solid, or oils, if liquid) are esters of glycerol and fatty acids.

$$\begin{array}{c} CH_2.O.C.R \\ | \quad \parallel \\ R.C.O.CH \quad O \\ \parallel \quad | \\ O \quad CH_2.O.C.R \\ \parallel \\ O \end{array}$$

A simple lipid   R represents a fatty acid residue

The fat represented here has three identical fatty acid residues; naturally occurring glycerides are invariably mixed glycerides containing two or three different fatty acid residues per molecule and up to 40 different residues per mole.

There are about 50 naturally occurring fatty acids found in the bound state; most of these are long, unbranched, aliphatic chains containing an even number of carbon atoms. The most abundant naturally occurring saturated fatty acids are:

palmitic acid (a C_{16} acid) $C_{15}H_{33}CO_2H$
stearic acid (a C_{18} acid) $C_{17}H_{35}CO_2H$

The most widely distributed unsaturated fatty acids (acids which contain one or more double bonds) are:

oleic acid (C_{18}) $\triangle\,^9 C_{17}H_{33}CO_2H$
linoleic acid (C_{18}) $\triangle\,^{9:12} C_{17}H_{31}CO_2H$
linolenic acid (C_{18}) $\triangle\,^{6:9:12} C_{17}H_{29}CO_2H$

Plants can synthesise all their constituent fatty acids, but animals are unable to synthesise highly unsaturated fatty acids; these acids have been shown to be essential to growth and 1% of fat intake must be in the form of highly unsaturated fatty acids. The degree of unsaturation of the constituent fatty acids influences the physical state of the glyceride; the greater the percentage of unsaturated fatty acids, the lower the melting point. For example, highly unsaturated glycerides are oils, and saturated glycerides, or fats containing a high proportion of saturated fatty acids, are solids. Glycerides are non-polar lipids and are soluble in non-polar solvents such as hydrocarbons, ether, acetone, etc.

Waxes are esters of fatty acids and alcohols other than glycerol.

Complex Lipids

Complex lipids are divided into phospholipids and sphingolipids.

PHOSPHOLIPIDS

Phospholipids or phosphatides are derived from α-glycerophosphoric acid.

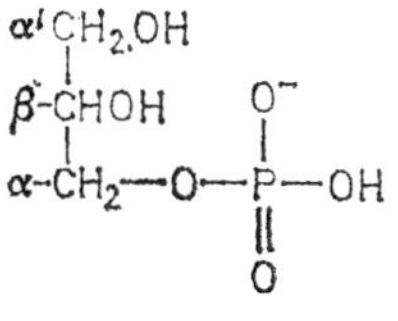

$$\begin{array}{l} \alpha'\text{CH}_2\text{OH} \\ \beta\text{-CHOH} \\ \alpha\text{-CH}_2\text{---O---}\underset{\underset{O}{\|}}{P}(\text{O}^-)\text{---OH} \end{array} \quad \alpha\text{-glycerophosphoric acid}$$

A more general formula of phospholipids is

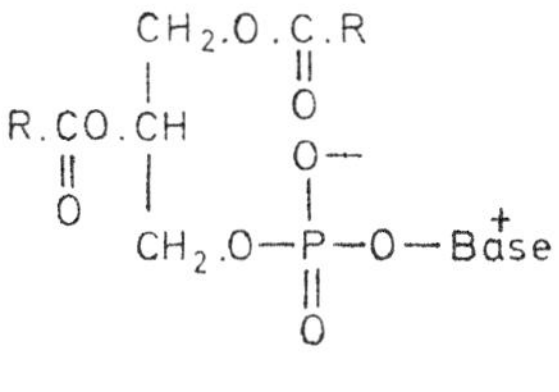

$$\begin{array}{l} \text{CH}_2.\text{O}.\text{C}.\text{R} \\ \phantom{\text{CH}_2.\text{O}.}\|\\ \phantom{\text{CH}_2.\text{O}.\text{C}.}\text{O} \\ \text{R}.\text{CO}.\text{CH} \\ \phantom{\text{R}.\text{C}}\|\phantom{\text{O}.\text{CH}}\text{O---}\\ \phantom{\text{R}.\text{C}}\text{O} \\ \phantom{\text{R}.\text{CO}.}\text{CH}_2.\text{O---}\underset{\underset{O}{\|}}{P}\text{---O---Base}^+ \end{array}$$

Formulas of bases generally present in phospholipids are given below.

	Name of lipid
(a) Base = choline	phosphatidyl choline

$$= HO.CH_2.CH_2.\overset{+}{N}(CH_3)_3 \text{ (trivial name is lecithin)}$$

(b) Base = ethanolamine — phosphatidyl ethanolamine (cephalin is a trivial name for a mixture of phosphatidyl ethanolamine and phosphatidyl serine)

$$= HO.CH_2.CH_2.\overset{+}{N}H_3$$

(c) Base = serine — phosphatidyl serine

$$= HO.CH_2.\underset{CO_2H}{CH}.\overset{+}{N}H_3$$

(d) Base = inositol — phosphatidyl inositol

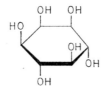

In addition, traces of phosphatidic acid (a hydrogen atom has been substituted for the base) exist in living organisms.

Phospholipids are polar compounds and are soluble in moderately polar solvents, such as methanol. They are widely distributed in animal and plant tissue, and are often associated with the cell membrane. Since they possess both a hydrophilic and a hydrophobic portion, they act as surface-active agents.

SPHINGOLIPIDS

Sphingolipids all contain the compound sphingosine.

$$CH_3(CH_2)_{12}.CH=CH.\underset{HO}{CH}.\underset{NH_2}{CH}.CO_2H$$

Sphingosine

They include sphingomyelin, cerebrosides and gangliosides, and are widely distributed in brain tissue.

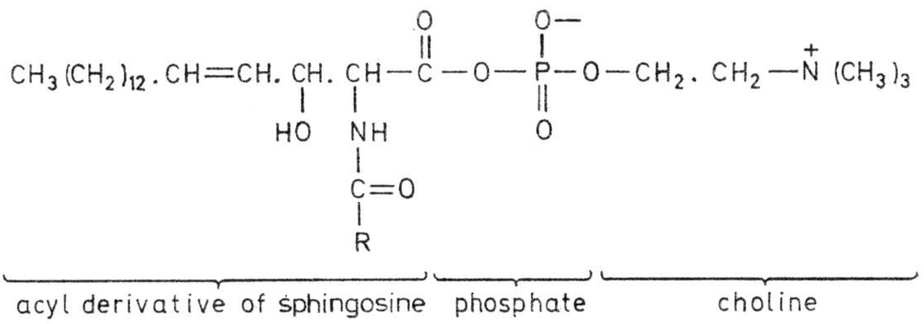

LIPIDS 101

$$CH_3(CH_2)_{12}.CH=CH.\underset{HO}{CH}.\underset{\underset{\underset{R}{C=O}}{NH}}{CH}-\overset{O}{\underset{\|}{C}}-O-\overset{O^-}{\underset{\underset{O}{\|}}{P}}-O-CH_2.CH_2-\overset{+}{N}(CH_3)_3$$

$\underbrace{\text{acyl derivative of sphingosine}}$ $\underbrace{\text{phosphate}}$ $\underbrace{\text{choline}}$

Sphingomyelin

Sterols

Sterols are derivatives of the saturated cyclopentanophenanthrene nucleus.

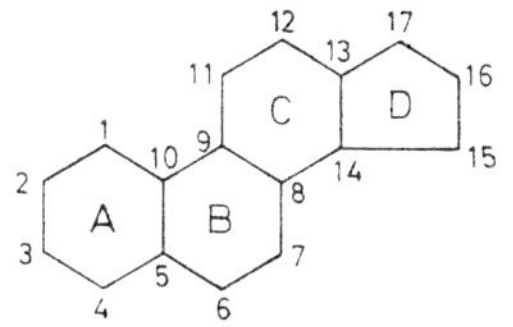

Saturated cyclopentanophenanthrene

Typical steroids contain a hydroxy group on C_3, a double bond on C_5, a methyl group on C_{10} and C_{13} and a side chain on C_{17}. Examples of common sterols are cholesterol (an animal sterol), sitosterol (a plant sterol), the bile acids and certain groups of hormones.

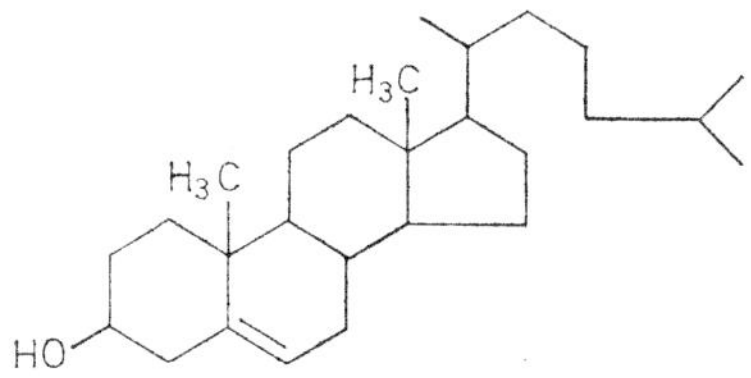

Cholesterol

Sterols also exist as esters, though they occur more abundantly in the free state. They are neutral lipids and are therefore soluble in neutral solvents. They display characteristic colours with concentrated sulphuric acid due to the resonance between the hydroxy group and the double bond.

SEPARATION

Successful separation of lipids based on solubility differences is extremely difficult. Nowadays the lipids are extracted with suitable solvents or a solvent mixture and then separated by chromatography. Small quantities of lipids are separated by thin layer chromatography (TLC) or reverse-phase chromatography. In the latter case the stationary phase is the non-polar phase, which is achieved by impregnating the paper with silicone, and the mobile phase is the aqueous phase. Larger quantities of lipids are often separated by silicic acid column chromatography.

Fatty acids are identified, separated and estimated by gas–liquid partition chromatography. In this type of chromatography the solute mixture to be separated must readily vaporise, and for this reason the fatty acids are converted to their methyl esters. The stationary phase is an involatile liquid which is supported in a column of inert material such as Celite. The solutes must be soluble or partially soluble in the stationary phase; and in the separation of fatty acids, high molecular weight esters, such as glycol adipate, are used. The mobile phase is an inert gas such as nitrogen, argon, helium, etc. This phase is often referred to as the carrier gas. The solute mixture is spotted, by means of a hypodermic needle, onto the top of a heated column. The solutes vaporise, and the rate at which they are eluted through the column is dependent on the rate at which they volatilise and on their solubility in the stationary phase. As the solutes are eluted from the column, they pass through a suitable detector; this is connected to a recorder which records the results. In general, the lower the molecular weight of the acid, the faster it is eluted from the column; and the greater the degree of saturation, the faster it is eluted from the column.

Fatty acids are identified by their retention time in the column —that is to say, by the length along the horizontal axis before they are recorded.

EXPERIMENT 24
PROPERTIES OF GLYCERIDES

Reagents

Olive oil
Lard
Lecithin

LIPIDS 103

Rancid oil or fat
Chloroform
Ethanol
Glacial acetic acid
Glycerol
Methanol
Anhydrous $KHSO_4$
Bromine in chloroform (about 1 ml Br_2 in 20 ml $CHCl_3$)
10% w/v KI
2 M $CaCl_2$ solution
2 M $MgCl_2$ solution
2 M $CuSO_4$ solution
1 : 1 methylene blue : oil-soluble yellow
Ethanolic NaOH (approx. 1 M)
Solid NaCl

Procedure

SOLUBILITY

Note the physical state of (1) olive oil (or another vegetable oil), (2) lard (or another animal fat) and (3) lecithin.
Test their solubilities in (a) water, (b) chloroform and (c) ethanol.

ACROLEIN TEST—TEST FOR GLYCEROL

Glycerol is dehydrated with anhydrous potassium hydrogen sulphate with the formation of an unsaturated aldehyde, acrolein.

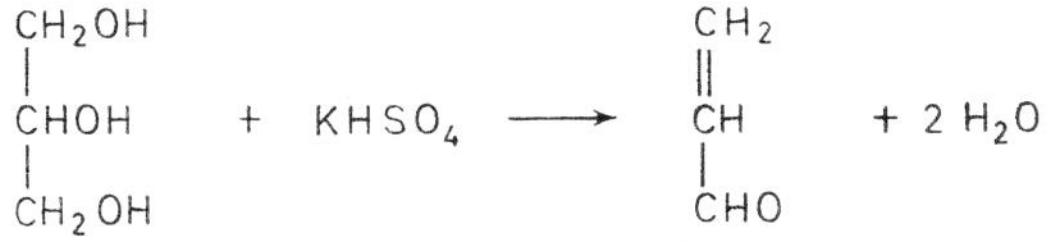

$$\begin{array}{c} CH_2OH \\ | \\ CHOH \\ | \\ CH_2OH \end{array} + KHSO_4 \longrightarrow \begin{array}{c} CH_2 \\ \| \\ CH \\ | \\ CHO \end{array} + 2H_2O$$

Acrolein

Acrolein has a characteristic unpleasant odour.

Add about three spatula-points (about 1·5 g) of anhydrous potassium hydrogen sulphate to a Pyrex test-tube and one drop of glycerol. Heat the tube strongly and carefully note the odour of the fumes evolved.

TEST FOR UNSATURATION

Add one drop of olive oil, one spatula-point of lard and one spatula-point of lecithin to separate dry test-tubes and dissolve the lipids in about 1 ml of chloroform or ethanol. Then add 1 ml of $CHCl_3$ to another test-tube to act as a blank. By means of a pasteur pipette, add dropwise a solution of bromine in chloroform until a definite yellow colour is produced. Note the number of drops required in each case and comment on the results.

TEST FOR PEROXIDES

Dissolve about 1 ml of olive oil in 1 ml of chloroform, add 2 ml of glacial acetic acid and one drop of 10% potassium iodide solution. Mix well and leave for 5 min. Repeat the experiment with rancid oil or fat.

The presence of peroxides is denoted by the liberation of iodine.

PREPARATION AND PROPERTIES OF SOAPS

Soaps are salts of fatty acids. Common everyday soaps are the sodium and potassium salts of naturally occurring fatty acids; these soaps are soluble. Other salts of naturally occurring fatty acids are insoluble, e.g. magnesium and calcium soaps, which are often referred to as 'scum'.

Soaps are surface-active agents. They stabilise emulsions and reduce the surface tension of the medium. Sodium and potassium soaps stabilise oil-in-water emulsions and other soaps stabilise water-in-oil emulsions. They are prepared by saponification (alkaline hydrolysis) of fats according to the equation:

$$\begin{array}{l} CH_2.O.\overset{\|}{\underset{O}{C}} R_1 \\ | \\ CH.O.\overset{\|}{\underset{O}{C}} R_2 \\ | \\ CH_2.O.\overset{\|}{\underset{O}{C}} R_3 \end{array} + 3\,NaOH \longrightarrow R_1CO_2Na + R_2CO_2Na + R_3CO_2Na + \begin{array}{l} CH_2OH \\ | \\ CHOH \\ | \\ CH_2OH \end{array}$$

Dissolve 2 ml of olive oil in 5 ml of benzene in a 100-ml round-bottomed flask and add 25 ml of approximately 1 M ethanolic NaOH. Fit the flask with a reflux condenser and reflux the mixture on a heated mantle for 30 min. Rearrange the apparatus for distillation and evaporate the mixture to dryness. Redissolve the residue in 50 ml of methanol : water 1 : 1 mixture, leaving the flask on the heated mantle to facilitate resolution. Transfer the mixture to a beaker and salt out the soaps by adding solid sodium chloride until no more dissolves. Filter off the precipitate of soaps which floats to the surface.

Dissolve a portion of the filtered soaps in 5 ml of water and add dilute solutions of (1) $CaCl_2$, (2) $MgSO_4$ and (3) $CuSO_4$ to aliquots of this solution. Comment on the results.

Dissolve another portion of soap in 10 ml of water. Place 10 ml of olive oil and 10 ml of soap solution into a separating funnel and mix well. Pour the emulsion into a petri dish containing a small amount of 1 : 1 methylene blue : oil-soluble yellow dye mixture.

Repeat this experiment using 10 ml of lecithin solution instead of soap solution.

Comment on the type of emulsions formed.

EXPERIMENT 25
ESTIMATION OF SAPONIFICATION NUMBER

The saponification number of a fat is defined as the number of mg of KOH required to saponify 1 g of fat. The value of the saponification number of a fat depends on the average molecular weight of its constituent fatty acids; the greater the average molecular weight, the smaller the number of potential carboxyl groups in a given weight of fat and therefore the smaller the saponification number.

Reagents

40% w/v fat solution in benzene
Benzene
Ethanolic NaOH (approx. 0·5 M)
Standardised HCl (approx. 0·5 M)
Phenolphthalein

Procedure

Pipette, in duplicate, 5 ml of the given fat solution into a 250-ml conical flask and add, by means of a pipette, 25 ml of the ethanolic NaOH solution. Into a similar flask pipette 5 ml of benzene and

25 ml of the ethanolic NaOH as a blank. Fit the flasks with reflux condensers and reflux the mixtures for 30 min. After refluxing, dismantle the condensers and add 50 ml of water to each flask, allowing the water to wash the corks and the inside and outside of the tips of the condensers. Titrate the alkali content of each flask with standardised approximately 0·5 M HCl, using phenolphthalein as indicator.

Calculation

Molarity of HCl $= y$ M

titre value (blank sample) $= b$ ml

1 mol HCl $\equiv$ 1 mol KOH

b ml (titre) $\times y$ M HCl $\equiv b \times y$ mmol KOH

$b \times y$ mmol KOH $\equiv$ 5 ml 40% w/v fat soln. $=$ 2g fat

$b \times y \times 56$ mg KOH $\equiv$ 2g fat

Saponification No. $= \dfrac{b \times y \times 56}{2}$

Also determine the average molecular weight of the constituent fatty acids of the fat.

1 mol KOH $\equiv \frac{1}{3}$ mol fat

$= 1$ mol fatty acid $+ \frac{1}{3}$ mol glycerol

$- 1$ mol water

$= 1$ mol fatty acid $+ 12\cdot7$

1 mol fatty acid $\equiv 1$ mol KOH $- 12\cdot7$
(average molecular weight)

$b \times y$ mmol KOH $\equiv$ 2g fat

1 mol KOH $\equiv \dfrac{2}{b \times y} \times 1000$

1 mol fatty acid $\equiv \left(\dfrac{2}{b \times y} \times 1000\right) - 12\cdot7$
(average molecular weight)

EXPERIMENT 26
EXTRACTION AND SEPARATION OF NEUTRAL LIPIDS

No single solvent is suitable for extracting all lipids present in natural products. Use of neutral solvents such as ether, acetone, etc., extracts glycerides, sterols and small amounts of complex lipids, whereas use of polar solvents such as ethanol extracts the majority of complex lipids. Chloroform/methanol mixture is a commonly used solvent mixture for extracting the majority of lipids. Extraction is often carried out in an atmosphere of nitrogen which prevents auto-oxidation of the unsaturated fatty acids.

In this experiment the neutral fats are extracted with ether. The sterols are separated from the glycerides by saponification, and the 'total neutral fat' fraction and the 'unsaponifiable fat' fraction are separated by thin layer chromatography (TLC).

Lipid mixtures are separated by TLC. The principle is the same as that of paper chromatography except that the aqueous stationary phase is absorbed in an inert medium such as silica, instead of paper. Also, the development is faster and the resolution is better. (Fats are very poorly separated by paper chromatography because of streaking, which is due to the absorption of lipids along the paper.) In this type of chromatography a glass plate is evenly covered with a thin layer (about 0·25–1·0 mm thickness) of an inert porous medium, usually silica gel or alumina. The plate is dried and then the solute mixture to be separated is spotted at one end of the chromatogram, which is then developed in a suitable solvent mixture. In TLC, when an inorganic material supports the stationary phase, the spots can be located with concentrated H_2SO_4 followed by charring.

Reagents and Apparatus

Diethyl ether
Chloroform
9 M H_2SO_4
Conc. H_2SO_4
Acetic anhydride
Ethanolic NaOH (approx. 1 M)
Solvent system   hexane : ether : acetic acid : methanol
  90 : 20 : 2 : 3 by volume

Silica gel
Walnuts
Soxhlet extraction apparatus
Electrically heated mantles
Chromatography tank
Glass plates
Spreader for thin layer plates
Calibrated capillary pipette
Quickfit apparatus

Procedure

EXTRACTION

Finely grind about 2 g of walnuts and place the nuts in a Soxhlet thimble. Plug the thimble with a small piece of cotton wool; this prevents pieces of nut from blocking up the extraction apparatus. Set up the Soxhlet apparatus as in Figure 4.1, half-fill the flask with ether and circulate water through the reflux condenser.

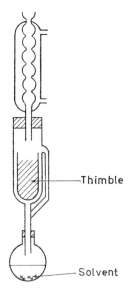

Figure 4.1. Soxhlet extraction apparatus

Extract the mixture for about 40 min, warming the flask gently on a heated mantle. Rearrange the apparatus for distillation and evaporate off the ether to dryness. Redissolve the residue in 5 ml of chloroform. Remove 0·5 ml of this solution into a Quickfit test-tube labelled 'total lipid extract' and store in the refrigerator for subsequent chromatography.

To the remainder of the fat solution add 10 ml of approximately 1 M ethanolic NaOH. Fit the flask with a reflux condenser and reflux the mixture for 30 min. Rearrange the apparatus for distillation and evaporate the mixture to dryness. Add 20 ml of ether to the flask, dissolve as much of the residue as possible and transfer this solution to a separating funnel. Then dissolve the remainder of the residue in about 20 ml of water and transfer this aqueous solution to the separating funnel. Thoroughly mix the two layers, releasing the pressure from time to time, and allow the two layers to settle. Run off the aqueous layer and collect the ethereal upper layer in a small conical flask. Evaporate off the ether in a beaker of warm water (there must be no flame on the same desk when ether is used) and redissolve the unsaponifiable lipid fraction in 2 ml of chloroform. Place 1 ml of this solution in a Quickfit testtube labelled 'unsaponifiable lipid extract' and store in the refrigerator for subsequent chromatography.

With regard to the remainder of the fraction, test for cholesterol (Lieberman–Burchard test) as follows: add a further 2 ml of $CHCl_3$, 1 ml of acetic anhydride and one drop of conc. H_2SO_4, and mix. A positive reaction is a purple colour which changes to green on standing.

THIN LAYER CHROMATOGRAPHY OF LIPID EXTRACTS

Throughly clean a 20×20 cm glass plate and rinse the surface with acetone; do not touch the surface—only the edges. Place 10 g of silica gel in a stoppered vessel and add 20 ml of water; thoroughly mix and evenly spread the slurry (layer thickness, 0·25 mm) with a suitable spreader. If a suitable spreader is not available, place two glass rods, slightly thicker than the plate, on either side of the plate, pour the slurry onto the plate and then run another glass rod along them.

Place about 50 ml of solvent in the chromatography tank. Seal the lid and allow the solvent to saturate the tank for about 20 min. Mark four origins at the top end of the plate, where the solvent will not reach, with a fine needle. On origins 1 and 2 spot, with a calibrated capillary pipette, 10 μl and 20 μl, respectively, of 'total lipid extract'. On origins 3 and 4 spot 10 μl and 20 μl of 'unsaponifiable lipid extract'. When making the spots, try to touch the silica layer as gently as possible, as a 'hole' in the layer causes the solvent

110 PRACTICAL BIOCHEMISTRY. AN INTRODUCTORY COURSE

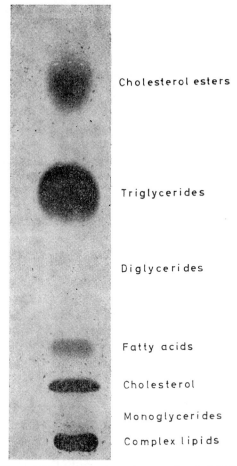

Figure 4.2. Thin layer chromatogram of neutral lipids

to run unevenly. Develop the chromatogram until the solvent approaches $\frac{1}{2}$ in from the top of the plate. Remove the plate from the tank, mark the solvent front and allow the excess solvent to evaporate in the air. Spray the plate with 9 M H_2SO_4 in the fume cupboard and heat in an oven at 110°C for 15 min to develop the spots. Determine the R_f values of the major spots. A chromatogram of neutral lipids is shown in Figure 4.2.

Sterols and sterol esters initially turn purple with concentrated H_2SO_4 before charring, so that their position can be determined during charring.

CHAPTER 5

Enzymes

THE ROLE OF ENZYMES AS CATALYSTS

The major task of metabolism is to provide energy for the maintenance of life. This is achieved by the breakdown of compounds of high potential energy to compounds of low potential energy. Conditions of metabolic reactions are controlled by the physiological environment—that is, the temperature is $37\pm1°C$ for mammalian metabolism and may vary slightly more widely for plant metabolism; the pH is in the region of neutrality; and no corrosive or poisonous reagents are present. Most of the chemical reactions which occur in metabolism can be carried out in a test-tube but the conditions required are much more extreme. For example, sugars are oxidised to CO_2 and H_2O at elevated temperatures with the use of a concentrated oxidising acid. The difference in conditions required for chemical and biochemical reactions is explained by the presence of enzymes. Enzymes are proteins which catalyse biochemical reactions. Enzymes, unlike chemical catalysts, are inactivated (denatured) during the reaction.

Consider the energy diagram in Figure 5.1 for the catalysed and uncatalysed conversion of $A \rightarrow B$. A is at a higher energy level than B and should therefore be spontaneously converted into B. However, before the reaction can proceed, the reactant must attain the energy of activation. The reaction temperature describes the average kinetic energy of the molecules. There will be many molecules possessing a higher and a lower energy state than the average. Molecules which possess the average kinetic energy plus the energy of activation will be spontaneously converted into B. If the energy of activation is high, a very small proportion, if any, of the molecules will possess the requisite energy and the reaction

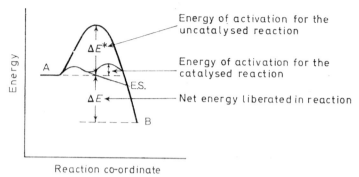

Figure 5.1. *Energy of activation of a catalysed an uncatalysed reaction*

will be slow, or non-existent, unless an additional energy input is applied. Enzymes facilitate a reaction by lowering the energy of activation so that a larger proportion of reactant molecules possess the requisite energy. One enzyme may catalyse the reaction in both directions. For the back-reaction, molecules of B must possess the average energy *plus* the liberated energy (ΔE) *plus* the energy of activation. One of the factors controlling the rate of the back-reaction depends on the energy difference between A and B (ΔE). The larger ΔE, the slower the back-reaction. Most 'reversible' reactions in biology have the reverse reaction catalysed by a different enzyme.

MECHANISM OF ENZYME REACTIONS

Most enzymes are highly specific. They are active for only one substrate or for one bond in a group of substrates. Furthermore, they may be stereospecific, in that they may be active for only a D- or L-isomer. There is direct evidence that in an enzyme-catalysed reaction, the enzyme and substrate form a complex—the enzyme-substrate complex. The enzyme possesses an active site onto which the substrate(s) are bound and where reaction occurs.

Enzyme kinetics is the study of the laws governing the rate of enzyme reactions.

In order to simplify the study of enzyme kinetics, only the initial reaction rate is considered, because, during the initial stages, the amount of product(s) formed is small, so the rate of the back-reaction can be ignored; and very little substrate has been used up, so the substrate concentration can be considered to be constant. The relationship of the rate of an enzyme reaction to substrate concentration is shown in Figure 5.2.

ENZYMES   113

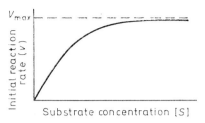

Figure 5.2. *Effect of substrate concentration on the rate of enzyme reaction (enzyme concentration constant)*

The rate of reaction increases with substrate concentration to a maximum. At the maximum reaction rate (V_{max}) all the enzyme is in the bound form—that is, in the form of the enzyme–substrate complex. When all the active sites of the enzyme are saturated with substrate, the substrate is present in stoichiometric excess of the enzyme. Under these conditions, the concentration of the enzyme is rate-limiting. In other words, where the substrate concentration is in excess, the rate of an enzyme reaction is directly proportional to the enzyme concentration. This relationship is shown in Figure 5.3.

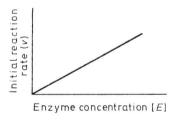

Figure 5.3. *Effect of enzyme concentration on the rate of enzyme reaction*

For this reason, enzyme assays are carried out using initial reaction rates in the presence of excess substrate. The formulation of the laws governing the rate of enzyme reactions where the substrate is not in excess is difficult. This is unfortunate, because the substrate, *in vivo*, may well be present in very small concentrations.

Co-factors

Many enzymes require a non-protein, thermostable, organic compound in order to function. These are either co-enzymes or prosthetic groups. A co-enzyme is a compound loosely attached to an enzyme, whereas a prosthetic group is a compound firmly bound to an enzyme (or another protein). Thus the difference between these two factors is not in *function* but in *degree*.

The function of a co-enzyme or a prosthetic group is to act as a carrier. They often assist in the removal of one of the products from a reaction, such as the removal of electrons or hydrogen in biological oxidations (for example, NAD, NADP, FAD, FMN, etc.), or the removal of CO_2 in decarboxylation reactions (for example, thiamine pyrophosphate, etc.). They also transport an active group to a reactant, as in the acetylation of fatty acids. Co-enzymes and prosthetic groups are usually less specific than enzymes in that they are specific for a certain type of reaction rather than a certain substrate. The majority of the B-vitamins are constituents of co-enzymes.

Many enzymes also require inorganic ions. Divalent cations, particularly Mg^{2+} and Zn^{2+}, are frequently required.

Temperature

The rate of an enzyme-catalysed reaction, as in the case of all chemical reactions, increases with temperature; but enzymes are proteins, and the rate of denaturation also increases with temperature. The rate of the reaction therefore increases at first and then decreases with rise in temperature (see Figure 5.4). The temperature at which the maximum rate occurs is referred to as the optimum temperature.

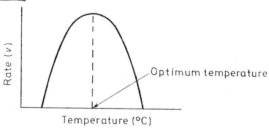

Figure 5.4. Effect of temperature on the rate of enzyme reaction

pH

Enzymes may be active only over a narrow pH range. They are not active over the whole pH range in which a protein remains in its native state. It is thought that the stereo-configuration of the active site is altered with small changes in H^+ ion concentration due to the destruction or formation of ionic bonds near the active site. The optimum pH is the pH at which the rate is maximum (see Figure 5.5).

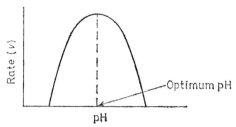

Figure 5.5. *Effect of* pH *on the rate of enzyme reaction*

ENZYME KINETICS

The first extensive kinetic studies on enzyme reactions were carried out by Michaelis and Menten (1913). Their method assumed that at equilibrium the rate of formation of enzyme-substrate complex equals the rate of disappearance of enzyme-substrate complex. The equation can be derived by the steps formulated below.

Consider the enzyme reaction

$$E_f + S \underset{k_{-1}}{\overset{k_1}{\rightleftharpoons}} ES \overset{k_2}{\longrightarrow} E + P$$

The decomposition of enzyme-substrate complex (ES) to enzyme plus products is a reversible reaction, but during the initial rate of the reaction, the amount of product(s) formed is negligible and the back-reaction can be ignored.

k_1 = velocity constant for formation of ES
k_{-1} = velocity constant for decomposition of ES into $E + S$
k_2 = velocity constant for decomposition of ES into $E + P$
[E] = total enzyme concentration
[E_f] = free enzyme concentration
 = [E] − [ES]
 (free enzyme concentration equals total enzyme [E] minus enzyme bound in enzyme-substrate complex [ES].)
[S] = total substrate concentration
 ≃ free substrate concentration
 (Amount of S combined with E is very small compared with amount of free S, and during the initial reaction the amount of S converted into P is negligible.) This is a major assumption in enzyme kinetics and in the derivation of the Michaelis-Menten equation.

At equilibrium: rate of formation of enzyme–substrate complex
= rate of disappearance of enzyme–substrate complex

Rate of formation of ES $= k_1[E_f][S]$

$= k_1([E]-[ES]).[S]$

Rate of disappearance of ES $= k_{-1}[ES]+k_2[ES]$

$= (k_{-1}+k_2).[ES]$

At equilibrium:

$$k_1([E]-[ES]).[S] = (k_{-1}+k_2).[ES]$$

$$\frac{([E]-[ES]).[S]}{[ES]} = \frac{(k_{-1}+k_2)}{k_1}$$

K_M (Michaelis constant) $= \dfrac{k_{-1}+k_2}{k_1}$

$$K_M = \frac{([E]-[ES]).[S]}{[ES]}$$

Rearranging the equation so that the ES terms are on the same side of the equation, we obtain

$$K_M.[ES]+[ES].[S] = [E].[S]$$

$$[ES] = \frac{[E].[S]}{K_M+[S]} \qquad (5.1)$$

The initial rate of reaction (v) is given by

$$v = k_2[ES] \qquad (5.2)$$

Substituting this value of [ES] in Equation (5.1), we obtain

$$v = \frac{k_2.[E][S]}{K_M+[S]} \qquad (5.3)$$

From Equation (5.3) it can be seen that the rate is proportional to the enzyme concentration.

When [S] is small and $K_M \gg [S]$, then $K_M+[S] \simeq K_M$. In this case the rate is proportional to the product of the enzyme concentration and substrate concentration.

When [S] is large and $K_M \ll [S]$, then $K_M+[S] \simeq [S]$. In this case the rate is independent of the substrate concentration.

Maximum rate (V_{max}) occurs when all the enzyme present is in the form of the enzyme–substrate complex ([E] = [ES]):

$$V_{max} = k_2[E]$$

$$k_2 = \frac{V_{max}}{[E]} \qquad (5.4)$$

ENZYMES   117

Substituting this value of k_2 in Equation (5.3), we obtain:

$$v = \frac{V_{max} \cdot [S]}{K_M + [S]} \qquad (5.5)$$

This is the Michaelis–Menten equation.
If we arrange Equation (5.5) in terms of K_M, we obtain

$$K_M = \frac{[S] \cdot (V_{max} - v)}{v}$$

$$K_M = [S] \cdot \left(\frac{V_{max}}{v} - 1\right) \qquad (5.6)$$

Thus, when the rate is half the maximum rate, the Michaelis constant is equal to the substrate concentration:

$$K_M = [S] \quad \text{when} \quad v = \frac{V_{max}}{2}$$

Where, in practice, the substrate concentration is required in excess, a concentration equal to $10K_M$ is usually employed.

The Michaelis constant can be obtained from a graph of rate plotted against substrate concentration (see Figure 5.6). The graph is called the Michaelis–Menten plot.

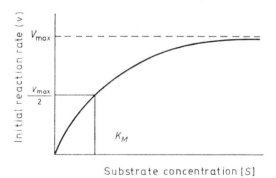

Substrate concentration [S]

Figure 5.6. The Michaelis–Menten plot

K_M and V_{max} are important constants in enzyme reactions. K_M indicates the stability of the enzyme–substrate complex. The value of K_M is inversely proportional to the affinity of the enzyme for the substrate. V_{max} is directly proportional to enzyme concentration and gives a measure of the rate of conversion of the enzyme–substrate complex to enzyme plus products.

A non-linear relationship is more difficult to obtain accurately from experimental results than a linear relationship. The Michaelis–Menten equation is therefore rearranged in the following way to give a linear equation:

$$v = \frac{V_{max} \cdot [S]}{K_M + [S]} \quad \text{(Michaelis–Menten equation)}$$

Taking reciprocals on both sides of the equation, we obtain

$$\frac{1}{v} = \frac{K_M + [S]}{V_{max} \cdot [S]}$$

$$\frac{1}{v} = \frac{K_M}{V_{max}} \cdot \frac{1}{[S]} + \frac{1}{V_{max}} \tag{5.8}$$

This is the straight line form of the Michaelis–Menten equation. It is usually referred to as the Lineweaver–Burk plot.

A plot of $1/v$ against $1/[S]$ gives a straight line of gradient K_M/V_{max} and intercept $1/V_{max}$ (see Figure 5.7).

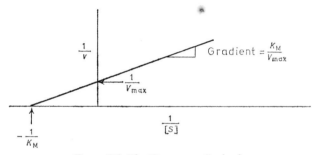

Figure 5.7. The Lineweaver–Burk plot

ENZYME ASSAYS

The purification of an enzyme consists essentially of isolating the protein with enzyme activity from a large number of other proteins. The methods used for the separation of proteins are used for the extraction and purification of enzymes. Many enzymes are difficult to purify and most of the assays in this chapter are carried out with crude enzyme suspensions.

Enzymes are assayed by their catalytic activity rather than the amount of enzyme protein present. Thus enzymes can be determined in the presence of other proteins. In practice, enzymes are assayed by determining the rate of appearance of products, or disappearance of substrate.

These assays are of two types.

The continuous assay. The enzyme and substrate are mixed together and some function dependent on the concentration of

product or substrate, such as gas pressure or absorbance, is continuously or periodically noted.
The discontinuous or discrete assay. This is further subdivided into two types.

1. The enzyme and substrate are mixed together for a fixed period of time, the reaction is stopped and the amount of substrate used or products formed is measured. This type of assay is referred to as the 'fixed time' or 'end-point' assay.
2. The enzyme and substrate are mixed together and the time is noted for a certain change, dependent on, say, the complete disappearance of substrate or utilisation of co-enzyme, to occur. An example is the time taken for the decolouring of a certain amount of methylene blue, which can sometimes act as an artificial co-enzyme. This method is inaccurate but has the advantage that it is simple to perform.

It is important in these assays to carry out adequate controls. A reagent blank containing all the reagents except the enzyme should be run, or, better still, a blank containing all the reagents together with inactivated enzyme. Each measurement should be obtained in duplicate or triplicate.

Many enzymes are inhibited by traces of heavy metal ions. It is therefore preferable to rinse all glassware with de-ionised water. Small amounts of cysteine, which reacts with heavy metal ions, or EDTA (ethylene diamine tetra-acetic acid), which chelates heavy metal ions, may be added as a precaution against inhibition from this source.

Units

TOTAL ACTIVITY

Many workers who originally worked on an enzyme designed their own units, and a variety of units appeared for a single enzyme. An International Commission has suggested a system of units to be used in enzyme assays. Their units are referred to as International Enzyme Units.

An International Enzyme Unit is defined as the amount of enzyme which catalyses the conversion of 1 μmol of substrate per minute. (The temperature and pH of the assay are stated—these are usually 30°C and the optimum pH.)

Total activity = μmoles of substrate used or μmoles of products formed per minute

SPECIFIC ACTIVITY

Specific activity = μmoles of substrate used or μmoles of products formed per minute per milligramme of protein

Specific activity × mg protein = total activity

ENZYME NOMENCLATURE

Enzymes are named according to the type of reaction they catalyse and the name(s) of the substrate upon which they act.

The International Commission on Enzymes in 1964 set out certain rules for enzyme nomenclature. It recommended that there be two names for enzymes: one *systematic* and one *working* or *trivial*. The *systematic* name of an enzyme, which affords precise identification, is formed in accordance with definite rules. The *trivial* name is sufficiently short for general use but not necessarily very exact; in many cases it is the name already in use. It is also recommended that names of enzymes ending in '-ase' be used only for single enzymes; in reactions which require more than one enzyme, the word 'system' should be included in the name.

Enzymes can be divided into six main groups based on the type of reaction they catalyse. The names of the groups and the type of reaction the enzymes catalyse are listed below.

Name of group of enzymes	Type of reaction the enzymes catalyse
Hydrolases	Hydrolysis
Oxidoreductases	Oxidation or reduction
Transferases	Transfer of a group from one substrate to another
Lyases	Non-hydrolytic addition or removal of groups
Isomerases	Isomerisation
Ligases (synthetases)	Joining of two molecules together coupled with the breakdown of a pyrophosphate bond in ATP or a similar triphosphate

Some examples of enzyme nomenclature follow.

Reaction	Trivial name	Systematic name
Triglyceride + H_2O = diglyceride + fatty acid	Lipase	Glycerol-ester hydrolase
L-lactate + NAD = pyruvate + reduced NAD	Lactate dehydrogenase	L-lactate: NAD oxidoreductase
ATP + D-hexose = ADP + D-hexose-6-phosphate	Hexokinase	ATP: D-hexose-6-phosphotransferase

INHIBITORS

There are various reagents which inhibit enzyme action. They can roughly be divided into the following groups.

Non-specific Inhibitors

Enzymes are proteins and any agent which alters the native state of a protein will tend to inactivate an enzyme. Examples are heat, extremes in pH, organic solvents, protein-precipitating agents such as trichloracetic acid, picric acid, sulphosalicylic acid, etc.

Inhibitors Acting at a Specific Site

Compounds which react with a functional group of an amino acid present at the active site of the enzyme will inactivate the enzyme (provided that the compound has access to this group in the protein). Reaction with the sulphydryl group (SH) (from the amino acid cysteine) has been studied more extensively than reactions with other groups. Many compounds react with the SH group to form either a disulphide or a metal mercaptide.

Examples are:

$$2R-SH \xrightarrow[-2H]{(Fe(CN)_6)^{3-}} R-S-S-R$$
disulphide

$$2R-SH \xrightarrow[-2H^+]{Hg^{2+}} R-S-Hg-S-R$$
mercuric mercaptide

(The SH groups in proteins can be protected by a small quantity of mercaptoethanol (C_2H_5SH) or dithiothreol.)

Inhibitors Reacting with Co-factors

Any compound which reacts with a co-factor will inhibit the enzyme system.
Cyanide forms a complex with Fe^{2+} and Fe^{3+} ions. Cyanide ions react with iron bound in the prosthetic group of cytochrome oxidase, thereby inhibiting cellular respiration. Similarly, fluoride ions inhibit magnesium-activated enzymes; oxalate ions inhibit calcium-activated enzymes, etc.

Competitive Inhibition

Structural analogues of substrates will inhibit enzyme activity, if they compete with the enzyme for binding at the active site. Enzyme–analogue complexes cannot, however, continue to form products. For example, malonic acid is a substrate analogue of succinic acid and competes with it for binding at the active site of succinic dehydrogenase.

$$\begin{array}{cc} CO_2H & CO_2H \\ | & | \\ CH_2 & CH_2 \\ | & | \\ CH_2 & CO_2H \\ | & \\ CO_2H & \\ \text{Succinic acid} & \text{Malonic acid} \end{array}$$

The degree of inhibition depends on the relative affinities of the substrate and inhibitor with the enzyme and on their concentrations. This type of inhibition is reversible.

EXPERIMENT 27
SALIVARY AMYLASE ACTIVITY

Salivary amylase catalyses the hydrolysis of starch to maltose and dextrins.
In this assay the time is noted for the complete disappearance of substrate. The substrate is starch and its presence can readily be detected with iodine solution.

Reagents and Apparatus

Saliva
1% w/v starch solution
1% w/v sodium chloride
Dil. iodine solution
Phosphate buffer at pH 6·6
Water-bath at 38°C
Stop-clock

Procedure

Pipette exactly 1 ml of unfiltered saliva into a 100-ml cylinder. Dilute to the 100-ml mark and mix well. Pipette 5 ml of 1% soluble starch solution into a test-tube, add 2 ml of 1% sodium chloride solution and 2 ml of buffer at pH 6·6. Mix well and place in a waterbath at $38 \pm 1°C$.
Prepare a series of 10 test-tubes, each containing 2 ml of pale yellow iodine solution.
Add 1 ml of diluted saliva to the starch mixture and return the tube to the water-bath immediately. Record the time of this addition. At the end of each minute, remove two drops of reaction mixture, using a teat pipette, and add to one of the tubes of iodine solution. Record the time when no change in colour appears in the iodine solution (achromic point). If the time is less than 5 min or more than 20 min, repeat the experiment using a different dilution of saliva to give an end-point in the region of 10 min.
Repeat the experiment using 2 ml of water instead of sodium chloride solution. Comment on the result.

Calculation

One amylase unit is defined as the amount of amylase which will catalyse the hydrolysis of 5 ml of 1% starch solution in 10 min under the conditions of the experiment.

$$\text{Amylase units} = 100 \, (\text{dilution factor}) \times \frac{10}{T} \text{ min}$$

T = reaction time for complete disappearance of starch

If the saliva was subsequently diluted, alter the dilution factor accordingly.

EXPERIMENT 28
SUCCINIC DEHYDROGENASE ACTIVITY

Dehydrogenases catalyse the transfer of hydrogen (or electrons) from the substrate to acceptor substances such as NAD, NADP or flavoproteins. Methylene blue and triphenyltetrazolium chloride act as artificial acceptor substances in place of flavoproteins. Both these compounds show a colour change on being reduced; methylene blue changes from blue in the oxidised form to colourless in the reduced form and the tetrazolium salt changes from colourless in the oxidised form to pink in the reduced form. Reduced methylene blue is readily reoxidised by atmospheric oxygen and therefore demonstration of the presence of dehydrogenases by use of methylene blue must be carried out in the absence of air. This is not the case for the tetrazolium salt, as the reduced form is not reoxidised by atmospheric oxygen. Succinic dehydrogenase is the only enzyme in the citric acid cycle to catalyse the transfer of electrons direct to a flavoprotein; the other dehydrogenases in the cycle catalyse the transfer of electrons to NAD or NADP and therefore, to demonstrate the presence of these enzymes, NAD or NADP is also required as a co-enzyme in the reaction mixture.

Reagents and Apparatus

Fresh rat muscle
0·25 M sucrose
0·05% w/v methylene blue
0·5% w/v triphenyltetrazolium chloride
0·1 M sodium succinate
0·5 M sodium malonate
50 mM phosphate buffer at pH 7
Homogeniser
Thunberg tubes
Water-bath at $37 \pm 1°C$

Procedure

METHYLENE BLUE METHOD

Homogenise 2–3 g of finely divided rat muscle in about 10–15 ml of 0·25 M sucrose. Keep the suspension in ice until it is required. Into three Thunberg tubes, labelled A, B and C, pipette the following reagents:

Tube	Methylene blue	Water	0·1M sodium succinate	Buffer pH 7
A	0·5 ml	0·5 ml	—	5 ml
B	0·5 ml	—	0·5 ml	5 ml
C	0·5 ml	—	0·5 ml	5 ml

Pipette 0·5 ml of the homogenate into the stopper of A and B and 0·5 ml of boiled homogenate into the stopper of C (see Figure 5.8). Grease the stoppers and insert them carefully into the tubes

Figure 5.8. A Thunberg tube

(without allowing any of the contents of the stopper to mix with the solution in the tube), so that the hole in the stopper coincides with the side arm of the tube. Evacuate the tubes on a suction pump for 3–4 min and then turn the stopper through 180° before dis-

126   PRACTICAL BIOCHEMISTRY. AN INTRODUCTORY COURSE

connecting the tube from the pump. Place the tubes in a water-bath at 37°C for about 3 min and then tilt the tubes to allow the enzyme suspension into the reaction mixture. Note this time for each tube. Continue to incubate the tubes and note the time taken to decolorise the methylene blue.

Repeat the experiment, adding 0·5 ml of 0·5 M sodium malonate to each tube. Comment on the result.

TETRAZOLIUM SALT METHOD

Into three test-tubes, labelled A, B and C, pipette the following reagents:

Tube	0·5% Tetrazolium salt	Water	0·1 M sodium succinate	Buffer pH 7
A	0·5 ml	0·5 ml	—	2 ml
B	0·5 ml	—	0·5 ml	2 ml
C	0·5 ml	—	0·5 ml	2 ml

Place the tubes in a water-bath at 37°C for 2–3 min. Add 0·5 ml of homogenate to tubes A and B and 0·5 ml of boiled homogenate to tube C. Record the time taken for the production of a definite pink colour.

EXPERIMENT 29

UREASE ACTIVITY

VARIATION OF ACTIVITY WITH TEMPERATURE AND pH

Urease is a plant enzyme which catalyses the hydrolysis of urea to ammonia, carbon dioxide and water. In this assay the ammonia produced is determined by aerating the liberated ammonia into saturated boric acid and then determining the amount of ammonium borate formed.

The experiment should be shared by the class, each student carrying out an assay at a different temperature or pH.

Reagents and Apparatus

0·1 M urea solution
Powdered soya bean, or urease solution containing 4 urease tablets
    in 100 ml water
Saturated boric acid

ENZYMES 127

Standardised HCl (approx. 0·01 M)
40% w/v sodium hydroxide
10% w/v trichloracetic acid
n-Octanol
Bromcresol green and methyl red indicators
Buffers at pH 4, 5, 6, 6·5, 7, 8 and 9
Water-baths at 30, 40, 50 and 60°C
(temperature tolerance, ±1°C)

Procedure

The apparatus consists of two boiling-tubes connected by glass tubing as shown in Figure 5.9. Boiling-tube 1 contains the reaction mixture and boiling-tube 2 contains saturated boric acid.

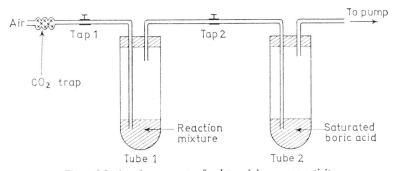

Figure 5.9. *Aeration apparatus for determining urease activity*

VARIATION WITH TEMPERATURE

Incubate, for exactly 30 min, the reaction mixture containing

(1) substrate 5 ml 0·1 M urea solution
(2) buffer   5 ml buffer at pH 6·5
(3) enzyme   2 spatula-tips (about 0·5 g) powdered soya bean or 5 ml urease solution

at
(a) 4°C   refrigerator
(b) 20°C   room temperature
(c) 30°C ⎫
(d) 40°C ⎬ water-baths
(e) 50°C ⎥
(f) 60°C ⎭

Time the reaction from addition of enzyme. During the incubation the longer glass tubing in tube 1 should be above the surface of the liquid and taps 1 and 2 should be closed. Include a reagent blank for each temperature containing all the reagents except the enzyme.

Remove the stopper from tube 1 and stop the reaction with 5 ml of 10% trichloracetic acid. Add 2–3 drops of octanol to the reaction mixture, to prevent frothing during aeration, followed by 5 ml of 40% NaOH. Add 4–8 drops of indicator (one part methyl red to three parts bromcresol green) to tube 2 containing the saturated boric acid. Adjust the glass tubing for aeration and open taps 1 and 2. Aerate the liberated ammonia into the boric acid for 20 min. Estimate the ammonia liberated by titrating the ammonium borate formed against standardised approximately 0·01 M HCl.

Plot enzyme activity (μmoles NH_3 formed per minute or titre value) against temperature.

Determine the optimum temperature of urease from the graph.

VARIATION OF ACTIVITY WITH pH

Incubate, for 30 min at 40°C, the reaction mixture containing

(1) substrate   10 ml 0·1 M urea
(2) enzyme   2 spatula-tips of soya bean or 5 ml of urease solution
(3) buffer   5 ml at pH 4·0, 5·0, 6·0, 6·5, 7·0, 8·0, 9·0

Time the reaction from addition of enzyme. Continue as before.

Plot enzyme activity (μmoles of NH_3 formed per minute or titre value) against pH.

Determine the optimum pH of urease from the graph.

Calculation

Let

$$\text{molarity of HCl} = y \text{ M}$$
$$\text{titre value} = b \text{ ml}$$
$$1 \text{ mol HCl} \equiv 1 \text{ mol } NH_3$$
$$b \text{ ml (titre)} \times y \text{ M HCl} \equiv b \times y \text{ mmol } NH_3$$
$$\equiv b \times y \times 1000 \text{ } \mu\text{mol } NH_3$$

EXPERIMENT 30
ALKALINE PHOSPHATASE ACTIVITY

Alkaline phosphatase catalyses the hydrolysis of phosphate esters in alkaline conditions (pH 8–10). In this experiment disodium p-Nitrophenyl phosphate is used as the substrate and serum is used as the enzyme solution.

$$NO_2\text{-}C_6H_4\text{-}O\text{-}P(=O)(ONa)_2 \xrightarrow[\text{pH 10 Mg}^{2+}]{\text{phosphatase}} NO_2\text{-}C_6H_4\text{-}OH + Na_2HPO_4$$

Disodium p-nitrophenyl phosphate → p-nitrophenol

This assay is an example of a 'fixed time assay'. The reaction mixture is incubated at 37°C for a fixed time, the reaction is stopped with caustic soda and the amount of product formed is determined. p-Nitrophenol, the product, is yellow and its concentration can be determined directly by measuring its absorbance at 415 nm (blue/purple filter).

The Michaelis–Menten and the Lineweaver–Burk plots are drawn for the variation of enzyme activity with substrate concentration. The Michaelis constant (K_M) and V_{max} are determined for the experiment.

Variation of activity with enzyme concentration is also investigated.

Reagents and Apparatus

Substrate	disodium p-nitrophenyl phosphate tetrahydrate: 10 μmol/ml (168 mg dissolved in 50 ml 0·001 M HCl)
Buffer at pH 10 containing Mg^{2+} ions	0·75 g glycine plus 21 mg $MgCl_2 \cdot 6H_2O$ dissolved in about 70 ml H_2O. Add 8·5 ml M NaOH and make solution up to 100 ml
Product	p-nitrophenol: 10 μmol/ml (69·5 mg dissolved in 50 ml 0·001 M HCl)

0·04 M NaOH
Serum
Water-bath at $37 \pm 1°C$
Colorimeter or spectrophotometer
Stop-clock

Procedure

VARIATION OF ACTIVITY WITH SUBSTRATE CONCENTRATION

Pipette, in duplicate, 0, 0·05, 0·1, 0·2, 0·4, 0·8 and 1·6 ml of substrate into a series of labelled test-tubes. Make the volumes up to 2 ml with water. Pipette 0·5 ml of buffer into each tube and incubate the tubes at 37°C for 5 min. Add, at 15-s intervals, 0·5 ml of serum to each tube, mix well and incubate the reaction mixtures for exactly 30 min. Stop the reaction by adding, with mixing, 3·5 ml of 0·04 M NaOH to each tube at 15-s intervals.

Read absorbance at 415 nm (blue/purple filter) against blank.

Prepare a standard curve of *p*-nitrophenol. Dilute the stock solution of *p*-nitrophenol 1 : 100 by pipetting 1 ml into a 100-ml volumetric flask and making up to the mark with 0·001 M HCl. Pipette, in duplicate, 0·1, 0·2, 0·4, 0·7 and 1·0 ml of diluted *p*-nitrophenol into a series of test-tubes. Make the volumes up to 6 ml with 0·04 M NaOH. Read the absorbance at 415 nm against water. Plot a standard curve of *p*-nitrophenol.

Read the amount of product formed from the standard curve.
Tabulate the results as follows:

Tube No.	Substrate concn., μmol/6 ml	Absorbance	Absorbance	Mean absorbance
1	0·5			
2	1·0			
3	2·0			
4	4·0			
5	8·0			
6	16·0			

Tube No.	μmol *p*-nitrophenol per 30 min per 0·5 ml serum (from standard curve)	S, mol/l	Activity (v), μmol per min per 1 serum	$\dfrac{1}{S}$, l.mol^{-1}	$\dfrac{1}{v}$
1					
2					
3					
4					
5					
6					

Plot a graph of:

1. Activity (μmol *p*-nitrophenol formed per min per 1 serum against substrate concentration (mol/l) (the Michaelis–Menten plot). Determine K_M and V_{max} from the graph (K_M = substrate concn. when $v = \frac{1}{2}V_{max}$).
2. The reciprocal of activity against the reciprocal of substrate concentration (the Lineweaver–Burk plot). Determine K_M and V_{max} from this graph.

VARIATION OF ACTIVITY WITH ENZYME CONCENTRATION

Pipette, in duplicate, 0·1, 0·25, 0·5, 1·0 and 1·5 ml of serum into a series of labelled test-tubes. Make the volumes up to 1·5 ml with water. Also prepare a blank containing 1 ml of serum and 0·5 ml of water. Pipette 0·5 ml of buffer to each tube and incubate the tubes at 37°C for 5 min. Add, at 15-s intervals, 1 ml of substrate to each tube and 1 ml of water to the blank, mix well, and incubate the reaction mixtures for exactly 15 min. Stop the reactions by adding 4 ml of 0·04 M NaOH to each tube at 15-s intervals.

Read absorbance at 415 nm against blank.

Read the amount of *p*-nitrophenol formed per 15 min from the standard curve.

Tabulate the results. Plot a graph of activity (μmol *p*-nitrophenol formed per minute) against enzyme concentration (volume of serum). Comment on the result.

EXPERIMENT 31
INVERTASE ACTIVITY

Invertase (β-fructofuranosidase) catalyses the hydrolysis of sucrose to 'invert sugar', an equimolar mixture of glucose and fructose.

Invertase is present in yeast. Yeast cells are disrupted by autolysis, the cell debris is removed by centrifugation and the supernatant provides cell-free yeast, which is a crude solution of invertase.

In this experiment enzyme activity is assayed by incubating the substrate and enzyme together for a fixed period of time. The reaction is then stopped and the amount of 'invert sugar' is estimated by use of dinitrosalicylic acid.

A suitable dilution of extract required for the estimation of activity is determined. By use of this dilution the total and specific activity for cell-free yeast is determined. The variation of activity with substrate concentration is investigated and K_M and V_{max} are calculated for this experiment.

Reagents and apparatus

Baker's yeast
0·1 M NaHCO$_3$
20 mM acetate buffer at pH 4·5
* 3,5-Dinitrosalicylic acid reagent
0·25 M sucrose
0·0025 M glucose, 0·0025 M fructose solution (0·0025 M invert sugar solution)
* Folin–Ciocaltea reagent
Solution A: 2% w/v Na$_2$CO$_3$ (anhydrous) in 0·1 M NaOH
Solution B: 0·5% w/v CuSO$_4$.5 H$_2$O in 1% sodium or potassium tartrate
Solution C: Mix 50 ml solution A with 1 ml solution B just before use
Standard protein solution containing 0·25 mg/ml
Homogeniser
Colorimeter or spectrophotometer
Thermostatically controlled water-bath at $25 \pm 1°C$

Procedure

PREPARATION OF CELL-FREE YEAST

Blend 50 g of yeast in 100 ml of 0·1 M NaHCO$_3$ for 1–2 min with a homogeniser. Transfer the slurry into a conical flask, plug the end of the flask with a piece of cotton wool and then place in a water-bath at 40°C for 24 h. After autolysis, centrifuge the mixture for 15 min. Decant off the clear supernatant into a measuring cylinder and record the volume. Store the cell-free extract in the refrigerator.

DETERMINATION OF DILUTION TO BE USED (DILUTION RANGE $2-10^{-3}$ ml OF ENZYME SOLUTION) FOR ENZYME ASSAY AND CALCULATION OF TOTAL ACTIVITY OF CELL-FREE YEAST

Prepare a 1 : 10, 1 : 100 and 1 : 1000 dilution of enzyme solution in a series of 100-ml volumetric flasks (for 1000-fold dilution pipette 1 ml of 1 : 10 dilution into a 100-ml volumetric flask and make up to the mark).

Into a series of test-tubes, pipette in duplicate: (a) 2 ml of enzyme solution; (b) 1 ml of enzyme solution; (c) 1 ml of 1 : 10 dilution; (d) 1 ml of 1 : 100 dilution; and (e) 1 ml of 1 : 1000 dilution. Make the volumes up to 2 ml with water. Also, for the blank use 2 ml of water. Add 1 ml of buffer to each tube and incubate the tubes at 25°C for 5 min. Add, at 15-s intervals, 1 ml of 0·25 M sucrose, mix well and incubate the reaction mixtures for exactly 15 min. Stop the reaction by adding 2 ml of 3,5-dinitrosalicylic acid reagent to each tube at 15-s intervals.

Develop the colour by heating the tubes in a boiling water-bath for 5 min. Cool and dilute the solutions to 10 ml with water.

Read absorbance at 540 nm (green filter) against the blank.

Prepare a standard curve of 'invert sugar'. Pipette, in duplicate, 0, 0·1, 0·4, 0·7 and 1·0 ml of 0·0025 M glucose, 0·0025 M fructose solution (0·0025 M invert sugar solution) into a series of test-tubes. Make the volumes up to 1 ml with water. Pipette 1 ml of 0·25 M sucrose followed by 2 ml of 3,5-dinitrosalicylic acid reagent into each tube. Continue the assay for invert sugar as above.

Plot a standard curve of invert sugar.

Read the amount of product formed per 15 min in the series of enzyme dilutions, from the standard curve. Tabulate the results. Note the dilution of enzyme solution which gives an absorbance reading in the range 0·2–0·8. This is the dilution to be used in the investigation of variation of activity with substrate concentration.

Calculate the total activity of cell-free yeast in terms of μmoles of product formed per minute.

DETERMINATION OF PROTEIN CONCENTRATION BY THE FOLIN–CIOCALTEA METHOD AND CALCULATION OF SPECIFIC ACTIVITY OF THE CELL-FREE YEAST

Carry out the protein estimation on a series of dilutions (dilution range 1 ml–1 × 10^{-2} ml of enzyme solution).

Pipette into a series of test-tubes 1 ml, 1 ml 1 : 10 dilution and 1 ml 1 : 100 dilution of enzyme solution. Prepare a blank containing 1 ml of water and a standard (in duplicate) containing 1 ml of standard protein solution. Add 5 ml of solution C to each tube, mix well and allow the tubes to stand at room temperature for 10 min. Add 0·5 ml of Folin–Ciocaltea reagent to each tube with immediate shaking. After 30 min, read absorbance at 750 nm (deep red filter) against the blank.

Calculate the protein concentration in the series of dilutions by comparison with the standard protein solution:

$$c_{unknown} = \frac{A_{unknown}}{A_{standard}} \times 0.25 \text{ mg protein}$$

(If all the dilutions of enzyme solution give absorbance values which are too high or too low, repeat the assay with another series of dilutions.)

Express protein concentration as mg protein in total extract and as mg protein per ml extract.

Calculate the specific activity of cell-free yeast in terms of μmol product formed per min per mg protein.

VARIATION OF ACTIVITY WITH SUBSTRATE CONCENTRATION

Pipette, in duplicate, 0, 0·05, 0·1, 0·2, 0·4, 0·8 and 1·6 ml of substrate into a series of labelled test-tubes. Make the volumes up to 2 ml with water. Pipette 1 ml of buffer into each tube and incubate the tubes at 25 °C for 5 min. Add, at 15-s intervals, 1 ml of a suitable constant dilution of enzyme solution to each tube, mix well and incubate reaction mixtures for exactly 15 min. Stop the reaction by adding 2 ml of 3,5-dinitrosalicylic acid reagent to each tube at timed intervals. Continue the assay for 'invert sugar' as above.

Read the amount of products formed per 15 min from the standard curve.

Tabulate the results as in Experiment 30.

Plot a graph of:

1. Activity (μmol invert sugar formed per min per ml extract) against substrate concentration (mol/l). Determine K_M and V_{max} from the graph.
2. The reciprocal of activity against the reciprocal of substrate concentration. Determine K_M and V_{max} from the graph.

EXPERIMENT 32
CATALASE ACTIVITY

Catalase is present in large amounts in liver. It can be extracted in water and then precipitated from a cold chloroform–methanol mixture.

The enzyme catalyses the breakdown of hydrogen peroxide to water and oxygen:

$$2 H_2O_2 \xrightarrow[\text{pH 7}]{\text{catalase}} 2 H_2O + O_2$$

The reaction is also catalysed by certain metal ions, particularly Fe^{3+} ions, and it is therefore advisable to use de-ionised water throughout this experiment.

The rate of the reaction is followed by continuously noting the amount of oxygen liberated with time. The amount of oxygen liberated is measured with an oxygen electrode.

An oxygen electrode (see p. 137) possesses a semipermeable membrane of 0·001-in Teflon, which is permeable mainly to molecular oxygen. This membrane encases a cell in which one electrode is platinum wire and the other is silver wire. A voltage of 0·6–0·8 V is applied across the terminals of the cell, which is insufficient to allow the passage of current through the cell (the resistance of the cell is too great); also, at this voltage only oxygen accepts electrons at the cathode. The cell is placed in the reaction mixture of catalase and hydrogen peroxide, and as the reaction proceeds O_2 is liberated and diffuses through the membrane. The O_2 in the cell accepts electrons at the Pt cathode to become OH^- ions; these ions conduct a current through the cell. The greater the concentration of OH^- ions, the greater the passage of current through the cell. The potential of the cell increases in direct proportion to the current; this is recorded by a pH meter which acts as a millivolt-meter.

The electrode is calibrated by removing the oxygen dissolved in water with sodium dithionite; this enables a reading on the scale to be directly correlated to a volume of oxygen.

Reagents and Apparatus

Ox liver
Methanol–chloroform mixture 2 : 1   The solvents are mixed in a conical flask surrounded by ice and stored in a refrigerator
20 mM phosphate buffer at pH 7
Hydrogen peroxide, 10-volume   Diluted 1 : 200
De-ionised water
Saturated KCl solution
Solid sodium dithionite
0·001-in Teflon
Pt wire
Ag wire
500-kΩ helipot (Beckman Instruments)
Two 10-kΩ fixed resistors
5-kΩ variable resistor

1·5-V dry cell
Switch
pH meter or millivolt-meter
Homogeniser
Magnetic stirrer

Procedure

PREPARATION OF CATALASE

Homogenise about 100 g of finely divided ox liver in 50 ml of de-ionised water. Transfer the homogenate to a 250-ml conical flask and add a further 50 ml of de-ionised water. Extract the enzyme in water by shaking the flask intermittently for 20 min. Add 50 ml of cold solvent mixture, vigorously shake for 30 s and immediately filter through Whatman No. 12 filter paper. Collect the filtrate in a conical flask surrounded by ice.

Place the filtrate, cooled in ice, in a refrigerator until the following day. During this time catalase will crystallise out. Centrifuge the crystals in a refrigerated centrifuge (if this is not available, filter at 4°C). Dissolve the crystals in 20 ml of de-ionised water at room temperature and store the enzyme solution in a refrigerator.

PREPARATION OF OXYGEN ELECTRODE AND CIRCUIT

Take a piece of soda glass tubing, about 5 cm in length, and heat one end in a Bunsen burner. Depress the heated end on an asbestos sheet so that the tubing does not quite close, and an opening of about 2 mm is left, e.g.

Take a piece of Pt wire (about 8 cm in length), encase this wire in plastic insulating tubing from an ordinary piece of electrical wiring, and leave the two ends uninsulated. Fit the open end of the glass tubing with a cork, bore a small hole through the centre to fit the insulated Pt wire and adjust the Pt electrode in the tubing so that an uninsulated tip is about 2 mm above the opening. Take a piece of Ag wire and bend the wire in a loop so that it can surround the Pt electrode inside the glass tubing. Place the straight end of the Ag wire through the cork and adjust the Ag electrode so that the loop is about 5–7 mm above the opening. Cut a piece of 0·001-in Teflon of area about $1 \times 1\cdot5$ in. Place the centre of the piece of Teflon over the small opening. Stretch the Teflon taut across the opening and then fix the membrane in position with an elastic band. Remove the electrodes from the tubing and fill the cell with saturated KCl solution. Replace the electrodes. An oxygen electrode is shown in Figure 5.10.

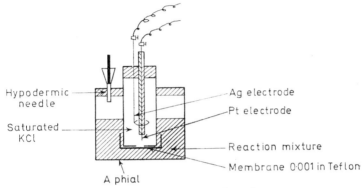

Figure 5.10. An oxygen electrode

Place the cell in a phial 1·5 cm in diameter (a test-tube with a flat bottom). Fit the phial with a cork and adjust the cell in the phial. Finally, fit a disposable hypodermic needle into the cork of the phial.

The circuit

Set up the circuit as in Figure 5.11.

CALIBRATION OF THE ELECTRODE

Connect the electrodes to the circuit by means of electrical screw clips. Remove the cell from the corked phial and place it in another phial three-quarters filled with aerated de-ionised water. Clamp the apparatus in position. Place the phial on a magnetic stirrer and place a magnetic flea in the phial. (A suitable flea is a piece of wire, about 0·5 cm in length, cut from a paper clip and sealed in glass

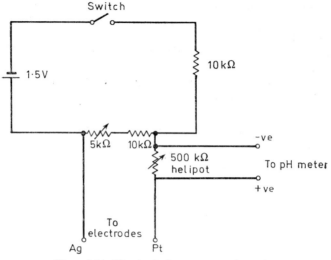

Figure 5.11. *The circuit for an oxygen electrode*

tubing). Adjust the resistance of both the variable resistor and the helipot so that the needle on the millivolt-meter or pH meter is roughly in the middle of the scale. Note the reading. Add a spatula-tip of sodium dithionite to the de-ionised water and stir the mixture continuously. Note the deflection of the needle. The decrease in the reading is equivalent to 0·23 μmol O_2/ml. (This is the amount of O_2 dissolved in water.)

N.B. Do not allow the Teflon membrane to remain in the dithionite solution for more than 2–3 min, for the dithionite will destroy the membrane in time.

Rinse the membrane with de-ionised water.

CATALASE ACTIVITY

Pipette 1 ml of substrate (H_2O_2, 10-volume, diluted 1 : 200) and 3 ml of buffer at pH 7, into the phial with a stopper. Stir the mixture with a magnetic stirrer. Place the electrodes in the phial. Adjust the resistances so that the needle on the pH meter is about in the middle of the scale. Note the reading. Using the hypodermic needle, add 50 μl of enzyme solution. Note the time of addition. Note readings on the scale in mV every 15 s for about 4 min. Tabulate (a) mV reading, (b) amount of oxygen liberated with time. Repeat the experiment.

Plot a graph of amount of oxygen liberated (μmol O_2/ml) against time.

N.B. Catalase is a very active enzyme and it is unnecessary to carry out the experiment at an elevated temperature.

EXPERIMENT 33
LIPASE ACTIVITY

Lipase catalyses the hydrolysis of glycerides to fatty acids and glycerol. The rate of the reaction is followed by noting the change of pH with time. Change of pH is due to the formation of fatty acids.

This method is not particularly accurate, because: (a) the activity of the enzyme changes with pH, and (b) the enzyme acts as a buffer and so inhibits changes of pH.

A more accurate method is continuous titration: the pH is kept constant with frequent additions of alkali and the rate of the reaction is followed by noting the titre values with time. This technique is usually carried out with an automatic apparatus, a pH-stat, which keeps the pH constant and at the same time plots a curve of titre values against time.

Lipase is present in large amounts in the pancreas, certain plants and bacteria. In this experiment it is extracted from broad beans.

Reagents and Apparatus

Broad beans
Olive oil
0·1 M HCl
0·1 M NaOH
Oleic acid
pH meter (preferably a meter with a ΔpH scale)
Magnetic stirrer

Procedure

EXTRACTION OF CRUDE LIPASE

Grind, with a pestle, about 20 g of shelled broad beans. Add 100 ml of water in 5-ml aliquots. On each addition, grind the mixture thoroughly until it is a smooth paste. Centrifuge (at about

3000 rev/min) for 20 min. Three layers are formed; pipette off the upper creamy layer and place in a stoppered conical flask. Discard the middle aqueous layer and re-extract the residue with a further 100 ml of water, proceeding as before. Again centrifuge for 20 min. Pipette off the upper layer and combine with the first portion. Label the suspension 'crude enzyme extract' and store in refrigerator.

LIPASE ACTIVITY

Prepare 40 ml of olive oil in water emulsion by adding 20 ml of olive oil, 18 ml of water, 2 ml of 0·1 M NaOH and 0·5 ml of oleic acid to a separating funnel, and shake vigorously (releasing pressure from time to time). An emulsion will form within a minute.

Standardise pH meter on ΔpH scale with buffer tablet at pH 6·99. Place electrodes in the emulsion and add dilute acid, dropwise with constant stirring, until pH is 7. Pipette 10 ml of emulsion to a 50-ml beaker. Place electrodes in substrate emulsion and add water until the tips of the electrodes are just covered; stir the mixture with a magnetic stirrer. Add 2 ml of enzyme suspension to the substrate, and record the time and pH. Tabulate changes of pH every 2 min for a total time of 60 min. Repeat the experiment using 2 ml of boiled enzyme suspension instead of enzyme suspension.

Plot a graph of lipase activity (change of pH (sample–blank)) against time.

APPENDIX I

pH

Acidity is due to an excess of hydrogen ions (H^+) over hydroxyl ions (OH^-) in solution. Alkalinity is due to an excess of hydroxyl ions over hydrogen ions in solution. A neutral solution is a solution which contains an equal concentration of H^+ ions and OH^- ions—for example, pure water.

The hydrogen ion concentration is a measure of acidity and alkalinity, but as this is usually very small it is more convenient to express acidity or alkalinity as the hydrogen ion exponent (represented as pH).

pH is defined as the negative logarithm to the base 10 of the hydrogen ion concentration (pH = $-\log_{10} [H^+]$).

A neutral solution has a pH of 7; this is determined from the hydrogen ion concentration of pure water. One mole of water is ionised in 10^7 litres at 25°C:

$$[H^+] = \frac{1}{10^7} \text{ g-ions/l}$$

$$\text{pH} = -\log_{10} 10^{-7}$$
$$= 7$$

Hydrogen and hydroxyl ion concentrations are related by the ionisation constant of water.

$$K_W = \frac{[H^+][OH^-]}{[H_2O]} \quad (K_W = \text{dissociation constant of water})$$

Since the degree of ionisation of water is very small, the concentration of water is considered constant.

$$K_W = [H^+][OH^-]$$

$$K_W = 10^{-7} \times 10^{-7}$$
$$= 10^{-14}$$

(K_W = ionisation constant of water; $[H^+] = [OH^-] = 10^{-7}$)

$$[H^+] = \frac{K_W}{[OH^-]}$$

$$pH = pK_W + \log_{10}[OH^-] \qquad (1)$$

$$pH = pK_W - pOH \qquad (2)$$

where $pK_W = -\log_{10} K_W$
$$= 14$$

and $pOH = -\log_{10}[OH^-]$

The pH scale ranges from 0 to 14; pH values < 7 indicate an acid solution and pH values > 7 indicate an alkaline solution. The further a pH value is from neutrality, the more acid or alkaline a solution is.

pH OF STRONG ACIDS AND STRONG BASES

Strong acids—for example, HCl—are completely ionised in solution and therefore the hydrogen ion concentration equals the concentration of the acid. ($[H^+]$ equals twice the acid concentration for a dibasic acid, etc.)

The pH of a molar monobasic strong acid solution is 0 and the solution increases by 1 pH unit for ecah tenfold dilution of acid.

Similarly, strong bases such as NaOH, are completely ionised in solution and therefore the hydroxyl ion concentration equals the concentration of base. The pH of a molar solution of a monobasic base is 14, and the pH decreases by 1 unit for each tenfold dilution.

pH OF WEAK ACIDS AND WEAK BASES

Weak acids and bases dissociate only to a limited extent even in very dilute solutions. The degree of dissociation and the hydrogen and hydroxyl ion concentration can be determined from their dissociation constants.

Consider the dissociation of a weak monobasic acid, $HX \rightleftharpoons H^+ + X^-$.

$$K_a = \frac{[H^+][X^-]}{[HX]}$$

where K_a = dissociation constant of weak acid, and $[H^+] = [X^-]$. [HX] represents the concentration of undissociated acid. As the degree of dissociation is small, [HX] is considered equal to total acid concentration.

$$K_a = \frac{[H^+]^2}{[acid]}$$

$$[H^+]^2 = K_a \cdot [acid]$$

$$pH = \tfrac{1}{2} pK_a - \tfrac{1}{2} \log_{10} [acid] \quad (pK_a = -\log_{10} K_a) \quad (3)$$

Equation (3) gives the pH of a solution of a weak monobasic acid. $pH = \tfrac{1}{2} pK_a$ for a molar solution of a monobasic weak acid and the pH value increases by 1 unit for each hundredfold dilution of acid.

Similarly, the pH of a solution of a weak base is given by

$$pH = 14 - \tfrac{1}{2} pK_b + \tfrac{1}{2} \log_{10} [base] \quad (4)$$

pH of a molar solution of a weak base $= 14 - \tfrac{1}{2} pK_b$, decreasing by 1 pH unit for every hundredfold dilution of base.

pH OF SALT SOLUTIONS

Ions of weak acids and weak bases react with water to form the undissociated acid or base. This alters the hydrogen ion concentration of water and often salt solutions do not possess a pH of 7.

Salts of a Strong Acid and Strong Base

These salt solutions, an example of which is NaCl solution, give a neutral reaction to litmus. None of the ions present reacts with water and therefore the hydrogen ion concentration of water is undisturbed.

Salts of a Strong Acid and Weak Base

These salt solutions, an example of which is NH_4Cl solution, give an acid reaction to litmus. The cation reacts with water to form the undissociated weak base; this causes an excess of H^+ ions in solution.

Consider a solution of a salt formed from a strong acid and a weak base, e.g. NH_4Cl solution:

$NH_4Cl \rightarrow NH_4^+ + Cl^-$ (The salt is completely ionised)

$NH_4^+ + H_2O \rightleftharpoons NH_4OH + H^+$ (Hydrolysis of the salt: $[NH_4OH] = [H^+]$)

$$K_H = \frac{[NH_4OH][H^+]}{[NH_4^+]}$$

(K_H = hydrolysis constant. H_2O is constant and therefore it is ignored)

$$K_H = \frac{K_W}{K_b} \quad \left(K_W = [H^+][OH^-]; \quad K_b = \frac{[NH_4^+][OH^-]}{[NH_4OH]}\right)$$

$$\frac{K_W}{K_b} = \frac{[H^+]^2}{[\text{salt}]} \quad ([NH_4^+] = [\text{salt}])$$

$$[H^+]^2 = \frac{K_W \cdot [\text{salt}]}{K_b}$$

$$\text{pH} = \tfrac{1}{2} pK_W - \tfrac{1}{2} pK_b - \tfrac{1}{2} \log_{10} [\text{salt}] \tag{5}$$

Equation (5) gives the pH of a solution of a salt formed from a strong acid and a weak base.

The pH of this type of salt solution increases by 1 pH unit for every hundredfold dilution of salt.

Salts of a Strong Base and Weak Acid

The salt solutions, an example of which is sodium acetate solution, give an alkaline reaction to litmus. The anion undergoes hydrolysis to form the undissociated weak acid; this causes an excess of OH^- ions in solution.

The pH of a solution of a salt formed from a weak acid and strong base is given by

$$\text{pH} = 7 + \tfrac{1}{2} pK_a + \tfrac{1}{2} \log_{10} [\text{salt}] \tag{6}$$

The pH of this type of salt solution decreases by 1 pH unit for every hundredfold dilution of salt.

Salts of a Weak Acid and a Weak Base

These salt solutions, an example of which is NH_4CN solution, may give a neutral, acid or alkaline reaction to litmus. Both the anion and cation undergo hydrolysis, the degree of which depends on the relative strengths of the acid and base from which the salt is derived. If $K_a = K_b$, the salt solution gives a neutral reaction; if $K_a > K_b$, the salt solution has a pH < 7, and conversely if $K_a < K_b$, the salt solution has a pH > 7.

The pH of a solution of a salt formed from a weak acid and weak base is given by

$$pH = \tfrac{1}{2} pK_W + \tfrac{1}{2} pK_a - \tfrac{1}{2} pK_b \qquad (7)$$

The pH of this type of salt solution is independent of concentration.

BUFFERS

A buffer is a solution which resists changes of pH on addition of an acid or an alkali. Buffers are mixtures; one constituent of the buffer absorbs hydrogen ions on addition of an acid and the other constituent absorbs hydroxyl ions on addition of an alkali.

The most common buffers are solutions of a weak acid and its sodium or potassium salt, or a weak base and the salt of that base and a strong acid. An example of the former is acetic acid and sodium acetate; and of the latter, ammonium hydroxide and ammonium chloride. Mixtures of two salts of the same acid, such as $H_2PO_4^-$ and HPO_4^{2-} are also considered examples of the first type; in this case the former (the salt which contains more replaceable hydrogen atoms) acts as the acid and the latter as the salt.

Consider the buffering action of a weak acid (HX) and its salt (NaX):

$HX \rightleftharpoons H^+ + X^-$ (Reaction 1: ionisation of acid)

$NaX \rightarrow Na^+ + X^-$ (Reaction 2: ionisation of salt)

The acid ionises only to a very limited extent and the salt is considered completely ionised.

In this case H^+ ions are absorbed by the salt and OH^- ions are absorbed by the acid.

Addition of H^+ ions causes Reaction 1 to 'go in the back direction'; the H^+ ions combine with the anions (X^-) from the salt to form the undissociated acid (HX) and, hence, the hydrogen ion concentration remains more or less constant. Conversely, addition of OH^- ions causes Reaction 1 to go in the forward direction; the OH^- ions combine with the H^+ ions to form water and again the hydrogen ion concentration remains more or less constant. (In buffers of a weak base and its salt, it is the base which buffers against H^+ ions and the salt which buffers against OH^- ions.)

The buffer constituent which absorbs H^+ ions is often referred to as the 'alkaline reserve' and the constituent which absorbs OH^- ions as the 'acid reserve'.

Buffers have a finite buffering capacity. When an equivalent amount of acid to the 'alkaline reserve' has been added, the buffering action is exhausted in the acid region. Conversely, when an equivalent amount of base to the 'acid reserve' has been added, the buffering action is exhausted in the alkaline region.

The pH of a buffer can be calculated from the dissociation constant of the weak acid (or base) and the concentrations of the acid (or base) and salt:

$$K_a = \frac{[H^+][X^-]}{[HX]}$$

where K_a = dissociation constant of acid.

Since the acid is ionised only to a very small extent, the concentration of the undissociated acid is considered equal to the total acid concentration $[HX] = [acid]$; and since the salt is completely ionised, the concentration of the anion is considered equal to the concentration of the salt $[X^-] = [salt]$. Then

$$K_a = \frac{[H^+][salt]}{[acid]}$$

$$[H^+] = K_a \cdot \frac{[acid]}{[salt]}$$

$$pH = pK_a + \log_{10} \frac{[salt]}{[acid]} \qquad (8)$$

Equation (8), which is known as the Henderson–Hasselbach equation, gives the pH range of a weak acid and its salt-type buffer.

Buffering action is greatest when $[salt] = [acid]$ (buffer possesses greatest buffering capacity in both directions). Then

$$pH = pK_a$$

In general, buffering action extends 1 pH unit in either direction from the pK_a value (ratio of salt to acid or vice versa at pH values of $pK_a \pm 1$ is about 91% : 9%):

The pH range of a weak acid and its salt-type buffer is given by
$$pH = pK_a \pm 1 \tag{9}$$
(For example, the pH range of acetic acid–sodium acetate buffer is $4 \cdot 7 \pm 1$. Therefore pK_a of HAc = $4 \cdot 7$.)
The pH range for a buffer of a weak base and its salt is given by
$$pH = (14 - pK_b) \pm 1 \tag{10}$$
For a list of buffers, see *Data for Biochemical Research*, ed. R.M.C. Dawson *et al.*, 2nd edn, pp. 483–506.

INDICATORS

Indicators are substances which change colour according to the hydrogen ion concentration of the solution. These indicators possess the properties of weak acids or bases.

The undissociated form of the indicator possesses one colour and the ionised form possesses another colour.

HI represents a weak-acid-type indicator which ionises to a limited extent:

$$\underset{\text{colour (a)}}{HI} \rightleftharpoons \underset{\text{colour (b)}}{H^+ + I^-}$$

where colour (a) = colour of undissociated indicator, and colour (b) = colour of anions I^-.

Addition of H^+ ions to the indicator solution causes the reaction to 'go in the back direction' with the formation of undissociated acid (HI). In this case very few I^- ions exist in solution and the colour of the indicator solution is colour (a). Conversely, addition of OH^- ions causes the reaction to go in the forward direction with the formation of anions I^- of colour (b).

The intermediate colour occurs when the intensity of colour (a) equals the intensity of colour (b); this is the neutralisation point of the indicator (its pK value).

The pH range of the indicator can be determined from the relative concentration of the two colour forms and the dissociation constant of the indicator (K_1):

$$K_1 = \frac{[H^+][I^-]}{[HI]}$$

$$= [H^+] \cdot \frac{\text{colour (b)}}{\text{colour (a)}}$$

$$[H^+] = K_1 \cdot \frac{\text{colour (a)}}{\text{colour (b)}}$$

$$pH = pK_1 + \log_{10} \frac{\text{colour (b)}}{\text{colour (a)}}$$

The neutralisation point occurs when colour (a) = colour (b); that is, when pH = pK_1.

The pH range in which an indicator changes colour is approximately given by

$$\text{pH} = \text{p}K_1 \pm 0\cdot 8 \qquad (11)$$

This range represents a change from 90% colour (a) to 10% colour (a) or vice versa.

If the indicator is strongly coloured in one form and hardly coloured in the other, the distinction of colour from about pK $\pm 0\cdot 5$ to pK $\pm 0\cdot 8$ on the highly coloured side is not distinguishable by eye. Hence the range is reduced (e.g. phenolphthalein).

The pH range of a weak-base-type indicator is given by:

$$\text{pH} = (14 - \text{p}K_1) \pm 0\cdot 8 \qquad (12)$$

For table of indicator ranges, see *Data for Biochemical Research,* ed. R. M. C. Dawson *et al.*, 2nd edn, p. 623.

MEASUREMENT OF pH

pH is most commonly measured with a glass electrode. The glass acts as a semipermeable membrane which is permeable only to H^+ and OH^- ions.

The glass electrode consists of a silver–silver chloride electrode in 0·1 M HCl (Figure A.1).

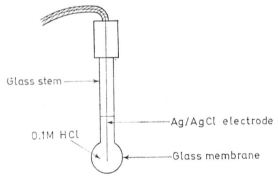

Figure A.1. A glass electrode

The potential produced by the glass electrode is dependent on the hydrogen ion concentration inside and outside the glass membrane and is given by

$$E_{\text{glass}} = E_{\text{Ag/AgCl}} + \frac{RT}{nF} \ln \frac{[H^+]_o}{[H^+]_i}$$

APPENDIX I. pH 149

where $[H^+]_o = [H^+]$ outside glass membrane;
$[H^+]_i = [H^+]$ inside glass membrane;
$E_{Ag/AgCl}$ = e.m.f. of standard Ag/AgCl/0·1 M HCl half-cell
= 0·283 V at 18°C;
R = gas constant = 8·314 J mol^{-1} K^{-1};
T = absolute temperature = 291K at 18°C;
n = charge on ions = 1; and
F = Faraday constant = 96 500 C mol^{-1}.

At 18°C

$$E_{glass} = 0·283 + 2·303 \frac{RT}{nF} (\log_{10} [H^+]_o - \log_{10} [H^+]_i)$$

$$E_{glass} = 0·283 - 0·058 \text{ pH} + 0·058 \text{ pH}$$

When pH = 1,

$$E_{glass} \text{ (at 18°C)} = (0·341 - 0·058 \text{ pH}) \text{ V}$$

The linearity of response of a glass electrode to pH depends largely on the composition of the glass. 'Soft' soda glass shows a linear response with pH up to about pH 10–11; in more alkaline conditions the OH$^-$ ions attack the glass and considerable errors occur. A great improvement has been achieved with lithium and alkaline earth glasses; these show a linear response up to about pH 13. At high pH's high concentrations of sodium ions also affect the electrode, but K$^+$ ions, etc., do not. Correction curves are available with most glass electrodes.

The glass electrode acts as a half-cell; when it is coupled with another half-cell, a potential difference is produced which can be measured. In the pH meter it is coupled to a standard reference half-cell, usually the calomel reference electrode.

The voltaic cell formed by the coupling of these two half-cells is given by:

$$\underbrace{Ag | AgCl | 0·1 \text{ M HCl} | Glass:}_{\text{glass electrode}} \begin{array}{c} \text{solution} \\ \text{of unknown} \\ \text{pH} \end{array} \| \underbrace{\begin{array}{c} \text{saturated} \\ \text{KCl} \end{array} | Hg_2Cl_2 | Hg}_{\text{calomel electrode}}$$

The e.m.f. (E) of the cell is given by

$$E = E_{ref.} - E_{glass}$$

where $E_{ref.}$ = e.m.f. of the reference calomel half-cell = +0·250 V at 18°C for saturated KCl.

$$E \text{ (at 18°C)} = 0·250 - (0·341 - 0·058 \text{ pH})$$
$$E = (-0·91 + 0·058 \text{ pH}) \text{ V}$$

The pH meter records the potential difference of the cell either directly as pH or as millivolts.

The potential difference of the cell is also temperature-dependent, the potential increasing by 0·2 mV/degC; for this reason accurate pH meters also possess temperature compensators.

APPENDIX II

Reagents

Acid Digestion Mixture

Dissolve 10 g of copper sulphate ($CuSO_4.5\ H_2O$) and 10 g of selenium dioxide in a mixture of 750 ml of concentrated H_2SO_4 and 250 ml of syrupy H_3PO_4.

Alkaline Copper Reagent

Dissolve 40 g of pure anhydrous sodium carbonate in about 400 ml of water. Add 7·5 g of tartaric acid and when the acid has dissolved, add 4·5 g of copper sulphate ($CuSO_4.5\ H_2O$) to the carbonate solution. Mix well and dilute to 1 litre. (If a sediment of cuprous oxide forms, filter the solution.)

Aniline Diphenylamine Reagent

Dissolve 4 ml of aniline, 4 g of diphenylamine and 20 ml of syrupy orthophosphoric acid in 200 ml of acetone. Prepare just before use.

Barfoed's Reagent

Dissolve 66 g of copper acetate and 10 ml of glacial acetic acid in water. Dilute to 1 litre.

APPENDIX II. REAGENTS

Benedict's Qualitative Reagent

Dissolve 73 g of sodium citrate and 100 g of anhydrous sodium carbonate in water. Dilute to about 850 ml. Dissolve 17·3 g of copper sulphate ($CuSO_4.5\ H_2O$) in about 100 ml of water and add this solution, with rapid stirring, to the citrate–carbonate solution. Dilute the reagent to 1 litre. (This reagent can be stored indefinitely.)

Benedict's Quantitative Reagent

Dissolve, by warming if necessary, 100 g of anhydrous sodium carbonate, 200 g of sodium or potassium citrate and 125 g of potassium thiocyanate in about 800 ml of water. Filter the mixture if necessary. Dissolve 18·0 g of copper sulphate ($CuSO_4.5\ H_2O$) in about 100 ml of water and add this solution, with constant stirring, to the first solution. Add 5 ml of a 5% w/v potassium ferrocyanide solution to the mixture and dilute the reagent to 1 litre. (Only the copper salt need be weighed exactly.)

Bial's Reagent

Dissolve 3·0 g of orcinol in 1 litre of concentrated HCl; add 3 ml of 1% w/v ferric chloride solution.

Biuret Reagent

Dissolve 1·5 g of copper sulphate ($CuSO_4.5\ H_2O$) and 6·0 g of sodium potassium tartrate (Rochelle salt, $NaKC_4H_4O_6.4\ H_2O$) in about 500 ml of water. Add, with constant stirring, 300 ml of 10% w/v sodium hydroxide solution (carbonate-free). Dilute the reagent to 1 litre. (Potassium iodide (1 g) may be added to the reagent if a deposit of cuprous oxide occurs.)

3,5-Dinitrosalicylic Acid Reagent

Dissolve, with warming, 10 g of 3,5-dinitrosalicylic acid in 200 ml of 2 M sodium hydroxide. Dissolve 300 g of Rochelle salt in about 500 ml of water and add this solution, with constant stirring, to the alkaline dinitrosalicylic acid solution. Dilute the reagent to 1 litre.

Fat-free milk

Reconstitute 1 pint of a commercial skimmed dried milk according to the directions given on the packet.

Fehling's No. 1 Solution

Dissolve 69·3 g of copper sulphate ($CuSO_4.5\ H_2O$) in water and dilute to 1 litre.

Folin–Ciocaltea Reagent

Dissolve 100 g of sodium tungstate ($Na_2WO_4.2\ H_2O$) and 25 g of sodium molybdate ($Na_2MoO_4.2\ H_2O$) in about 700 ml of water. Add 100 ml of concentrated HCl and 50 ml of 85% phosphoric acid to the mixture and reflux for 10 h in an all-glass apparatus. Add 150 g of lithium sulphate, 50 ml of water and a few drops of bromine. Boil the mixture for about 15 min to remove the excess bromine, cool, dilute to 1 litre and then filter. Store in the refrigerator. (The reagent should be golden-yellow in colour. If it acquires a greenish tint, it is unsatisfactory for use. The reagent may be regenerated by boiling with a few drops of Br_2.)

Dilute this concentrated reagent with two volumes of water just before use. Prepare in the fume-cupboard.

Foulger's Reagent

To 40 g of urea in 80 ml of 40% v/v H_2SO_4 add 2 g of stannous chloride. Boil the mixture until it is clear. Dilute the reagent to 100 ml with 40% v/v H_2SO_4.

Glucose Oxidase Reagent

Add 0·5 ml of Fermcozyme to about 80 ml of 0·15 M acetate buffer at pH 5. Dissolve 20 mg of peroxidase in 100 ml of 0·15 M acetate buffer at pH 5. Add 5 ml of this peroxidase solution to the Fermcozyme solution, mix well and then add 1 ml of 1% w/v tolidine in absolute ethanol. Make the reagent up to 100 ml with 0·15 M acetate buffer at pH 5 and store in a dark bottle in a refrigerator.

Fermcozyme is a stable liquid preparation of glucose oxidase containing 750 μg/ml. It is obtainable from Hughes and Hughes Ltd., 35 Crutched Friars, London E.C.3.

Peroxidase can be obtained from Courtin and Warner, Lewes, Sussex.

Gum Ghatti (0·2% w/v)

Fill a litre cylinder with water. Just below the surface, suspend a wire screen on which is placed 20 g soluble gum ghatti. Allow to stand for about 24 hours. Remove screen and filter mixture.

Hopkin–Cole Reagent

Place 10 g of powdered magnesium in a large conical flask. Add water until the magnesium is well covered. Slowly, with continuous shaking, add 250 ml of a cold saturated solution of oxalic acid to the flask. (Cool if necessary.) Filter, just acidify the filtrate with dilute acetic acid and dilute the reagent to 1 litre.

Millon's Reagent

Dissolve 10 g of mercury in 10 ml of fuming nitric acid. Add 40 ml of water and decant the clear solution after standing. Prepare in the fume-cupboard.

Nessler's Reagent

To 50 g of potassium iodide dissolved in 50 ml of water, add saturated mercuric chloride solution until a permanent precipitate just forms. Add 200 ml of 5 M NaOH (carbonate-free) and dilute to 1 litre. Decant the clear liquid after letting it stand.

Phosphomolybdic Reagent

Dissolve 70 g of molybdic acid and 10 g of sodium tungstate in 700 ml of 5% w/v sodium hydroxide solution. Boil for about 40 min to remove the ammonia in the molybdic acid. Cool and add 250 ml of 85% phosphoric acid. Dilute the reagent to 1 litre.

Seliwanoff's Reagent

Dissolve 0·5 g of resorcinol in 330 ml of concentrated HCl and dilute to 1 litre.

For buffers, see *Data for Biochemical Research*, ed. R. M. C. Dawson *et al.*, 2nd edn, pp. 483–506. Unless otherwise stated, use 100-mM buffers.

All concentrations are given as w/v unless otherwise stated. All solvents used for paper and thin layer chromatography are prepared by volume.

Bibliography

CANTARAW, A. and SCHEPARTZ, B., *Biochemistry*, 3rd edn, Saunders, Philadelphia (1962)

CLARK, J. M., *Experimental Biochemistry*, Freeman, San Francisco (1964)

COLE, S. W., *Practical Physiological Chemistry*, 9th edn, Heffer, Cambridge (1933)

CONN, E. E. and STUMF, P. K., *Outlines of Biochemistry*, 2nd edn, Wiley, New York (1966)

COUTTS, R. T. and SNAIL, G. A., *Polysaccharides, peptides and proteins*, Heinemann, London (1966)

DANIEL, LOUISE J. and NEAL, A. L., *Laboratory Experiments in Biochemistry*, Academic Press, New York (1967)

DAWSON, R. M. C. et al. (eds.), *Data for Biochemical Research*, 2nd edn, Oxford University Press, London (1969)

DIXON, M. and WEBB, E. C., *Enzymes*, 2nd edn, Academic Press, New York (1964)

FIESER, L. F. and FIESER, MARY, *Organic Chemistry*, 3rd edn, Reinhold, New York (1956)

HARROW, B. et al., *Laboratory Manual of Biochemistry*, 5th edn, Saunders, Philadelphia (1960)

LOCKWOOD, E. H., *Statistics: The How and the Why*, Murray, London (1969)

MCELROY, W. D., *Cell Physiology and Biochemistry*, 2nd edn, Prentice-Hall, Englewood Cliffs, N. J. (1964)

OSER, B. L. (ed.), *Hawk's Physiological Chemistry*, 14th edn, McGraw-Hill, New York (1965)

Recommendations 1964 for the International Union of Biochemistry, *Enzyme Nomenclature*, Elsevier, Amsterdam (1965)

SMITH, I. (ed.), *Chromatographie and Electrophoretic Techniques*, Vols. I and II, Heinemann, London (1960)

VARLEY, H., *Practical Clinical Biochemistry*, 4th edn, Heinemann, London (1967)

WEAST, R. C. (Editor-in-chief), *Handbook of Chemistry and Physics*, 47th edn, Chemical Rubber Publishing Co., Cleveland, Ohio (1966)

WEST, E. S. and TODD, W. R., *Textbook of Biochemistry*, 3rd edn, Macmillan, London (1966)

Index

Absorbance, 34
Absorption of light, 33
Acetals, 15
Acid reserve, 146
Acrolein test, 103
Aldonic acids, 22
Aldoses, 13
Alpha-helix, 59
Alkaline phosphatase, 129
Alkaline reserve, 146
Amino acids, 53
   table of, 54–55
   reactions of, 56–58
Amylase, salivary, 122
Amylopectin, 19
Amylose, 18
Aniline diphenylamine reagent, 51, 150
Arithmetic mean, 2
Asymmetric C-atoms, 14, 41

Barfoed's test, 22, 25
Barfoed's reagent, 150
Beer's law, 34
Benedict's test, 22, 24
   determination of sugars by, 30
Benedict's qualitative reagent, 151
Benedict's quantitative reagent, 151
Beta-helix, 60
Bial's test, 21, 26
Bial's reagent, 151
Biuret test, 73
   determination of proteins by, 90
Biuret reagent, 151
Buffer capacity, 146
Buffers, 145
   pH range of, 147

Calcium phosphate gel, 67
Calomel reference electrode, 149
Carbohydrates
   reactions of, 20–23
   separation of, 49
   structure of, 13–20
Casein, isolation of, 86
Catalase, 134
Cation exchange resin, 68, 76
Cell-free extract, 66
Cell-free yeast, 132
Cellulose ion-exchangers, 68
Cholesterol, 101
Chromatography
   ascending, 50
   descending, 50
   gas, 102
   gel filtration, 69, 96
   ion exchange, 67, 76
   paper, 49, 87
   thin layer, 102, 109
Co-enzymes, 113
Co-factors, 113
Collagen, 61
Colorimeter, 36
Coloured filters, 36
Complex lipids, 99

Dansyl chloride (DNS), 57
Denaturation of proteins, 63, 74
Dependent variable, 6
Deoxyhaemoglobin, 96
Dextrans, 20
Dextrins, 20
Dialysis, 76
3,5-Dinitrosalicylic acid, in determination of sugars, 47, 133
3,5-Dinitrosalicylic acid reagent, 151

INDEX

Dipeptides, 59
Disaccharides, 17
Dissociation constant, 143
Disulphide bridges, 62

Electrophoresis of amino acids, 79
Emulsions, 61, 104
Enzyme(s), 111–114
   kinetics, 115–118
   nomenclature, 120
   units, 119–120
Errors, 1
Ethylene glycol, 44
Exponential curves, 10

Fats, 98
Fatty acids, 98, 99
   test for peroxides, 104
   test for unsaturation, 104
Fearon's test, 25
1-Flouro-2,4-dinitrobenzene, 57
Folin–Ciocaltea method, determination of proteins by, 90, 133
Folin–Ciocaltea reagent, 152
Formic acid from periodate oxidation, 42–44
Formol titration, 58, 85
Foulger's test, 25
Foulger's reagent, 152
β-Fructofuranosidase, see Invertase
Fructose, 15
   periodate oxidation of, 43
Furan, 16
Furanose, 17
Furfural, 20

Gel filtration of haemoglobins, 95
Glass electrode, 148
Glucose, 15–17
   determination of, 27, 30, 37
   mutarotation of, 16, 41
   oxidation of, 22, 27, 43
Glucose oxidase test, 26
Glucose oxidase reagent, 152
Glyceraldehyde, 14
α-Glycerophosphoric acid, 99
Glycogen, 19
   in rat liver, determination of, 46
Graphs, 6–12
   linear, 8–12
   non-linear, 12
Gum ghatti, 94, 153

Helix, see Alpha-helix; Beta-helix
Hemiacetals, 15
Hemiketals, 15
Henderson–Hasselbach equation, 146
Hexoses, 14
Hopkin–Cole test, 72
Hopkin–Cole reagent, 153
Hydrogen bonding in proteins, 59
Hydroxy-methyl furfural, 20
Hyperbolic curves, 9

Independent variable, 6
Indicators, pH range of, 148
Inhibitors of enzymes, 121–122
Invertase, 131
Invert sugar, 29, 131
Iodine test for polysaccharides, 18, 19, 24
Ion exchange chromatography of amino acids, 76
Isoelectric point, of amino acids, 53–55
   determination of, 82
Isoelectric point precipitation of proteins, 64, 74

Ketals, 15
Ketoses, 13
Kjeldahl's determination of nitrogen, 91
K_M, see Michaelis constant

Lactose, 17
   in milk, determination of, 37
Lambert's law, 33
Lecithin, 100
Lieberman–Buchard test, 109
Lineweaver–Burk plot, 118
Lipase, 139
Lipids
   separation of, 102
   structure of, 98–101

Maximum enzyme reaction rate, 113, 116
Methaemoglobin, 96
Michaelis constant (K_M), 116, 117
Michaelis–Menton equation, 117
Millon's test, 71
Millon's reagent, 153

INDEX

Mobile phase, 49, 102
Molar extinction coefficient, 34
Molisch's test, 20, 24
Monosaccharides, 13–17
Mutarotation, 16, 41

Nelson's determination of sugars, 37, 38
Nessler's determination of nitrogen, 94
Nessler's reagent, 153
Ninhydrin, 56, 71
Non-reducing sugars, 18
Normal curve, 2–4

Observed rotation, 39
Oils, 98
Oligosaccharides, 13, 17
Optical density, *see* Absorbance
Osazone, 22, 26
Oxygen electrode, 135, 137
Oxyhaemoglobin, 96

Paper chromatography
  of amino acids, 87
  of sugars, 49
Partition chromatography, 49, 102
Pentoses, 14
Penultimate C-atom, 14
Periodate oxidation, 42
pH, 141–145
  effect of, on enzyme activity, 114
Phospholipids, 99–100
Phosphomolybdic acid reagent, 37, 38, 153
Photometry, 32–36
Picric acid, 65, 76
pK
  of amino acids, 82
  of weak acids and bases, 143
Polarimeter, 40
Polypeptides, 59
Polysaccharides, 13, 18–20
Precipitation of proteins, 64–65
Primary structure, 59
Probability, 3–4
Prosthetic group, 113
Proteins
  determination of, 90
  separation of, 65–69
  structure of, 58–63

Pyran, 16
Pyranose, 16

Quaternary structure, 63

Random coil, 61
Random errors, 2
Reducing sugars, 18, 22
R_f, 50
Ribose, 15

Sakaguchi's test, 72
Salting out
  of proteins, 64, 75
  of soaps, 105
Saponification, 104, 107
Saponification number, 105
Secondary structure, 59–62
Sephadex, 20, 69, 96
Seliwanoff's test, 20, 25
Seliwanoff's reagent, 154
Significance, 5
Significant figures, 6
Silica gel, 107, 109
Silver mirror test, 25
Simple lipids, 98
Soaps, 104
Solvent extraction, 64, 74
Specific activity, 120
Specific rotation, 39
Spectrophotometer, 36
Sphingolipids, 100
Sphingomyelin, 101
Standard deviation, 2
Standard error, 3
Starch, 18–19, 122
Stationary phase, 49, 102
Stereoisomers, 14
Sterols, 101
Succinic dehydrogenase, 122, 124
Sucrose, 18, 29
Sulphosalicylic acid, 65, 75
Sulphur test in proteins, 72
Systematic errors, 2

Temperature, effect of, on enzyme activity, 114
Tertiary structure, 63
Thin layer chromatography of neutral lipids, 107

INDEX

Tollen's test, 21
Total activity, 119
Transmittance, 34
Trichloracetic acid, 65, 75
Triphenyltetrazolium chloride, 124
Triple chain helix, 61

Ultracentrifugation, 66

Ultra-violet spectra of tyrosine, 89
Urease, 126
Uronic acids, 21

Xanthoproteic test, 71

Zwitterion, 53